3rd International Conference on Upstream Engineering and Flow Assurance 2015

Topical Conference at the 2015 AIChE Spring Meeting and 11th Global Congress on Process Safety

Austin, Texas, USA
26-30 April 2015

ISBN: 978-1-5108-0697-9

Printed from e-media with permission by:

Curran Associates, Inc.
57 Morehouse Lane
Red Hook, NY 12571

Some format issues inherent in the e-media version may also appear in this print version.

Printed by Curran Associates, Inc. (2015)

For permission requests, please contact AIChE
at the address below.

AIChE
3 Park Avenue
New York, NY 10016-5991

Phone: (203) 702-7660
Fax: (203) 775-5177

www.aiche.org

Additional copies of this publication are available from:

Curran Associates, Inc.
57 Morehouse Lane
Red Hook, NY 12571 USA
Phone: 845-758-0400
Fax: 845-758-2634
Email: curran@proceedings.com
Web: www.proceedings.com

TABLE OF CONTENTS

Chemical Engineering: Is This the Golden Age? 1
Marvin O. Schlanger

(17b) Adaption of Blosom (Blowout and Spill Occurrence Model) to High Performance Computing 2
Jason Vielma, Lawrence Sim, Kelly Rose, Rodrigo Duran-Sesin, Joe Umhoefer

(17c) Mechanism of Sloshing Minimization in Offshore Separator By CFD Simulation 3
Md. A Mahmud, Rafiqul Khan, Qiang Xu

(17d) Detailed Flow Modeling of Erosion in Subsea Equipment 4
Subodh Deodhar, Thomas Esch, Vedanth Srinivasan, L. Srinivasa Mohan

(37b) Erosion Model Development for High Concentration Slurry Flow Applications 5
Kuochen Tsai

(37c) State of Utah Energy Research Triangle (University of Utah, Brigham Young University, Utah State University) Project for Characterization of Uintah Basin Waxy Crude Oil Deposition in Pipelines 6
Richard Roehner

(37d) The Effect of Computational Fluid Dynamics Methodology on Solid Particle Erosion Predictions 7
Deval Pandya, Ronnie Russell, Brian Dennis

(69a) Evaluating Wax Yielding with a New Model 12
Rama Venkatesan, Krishnaraj Sambath, David Lavenson

(69b) A Self-Sufficient Wax Deposition Model for Horizontal Gas-Oil Slug Flow 13
Ekarit Panacharoensawad, Ake Rittirong, Cem Sarica

(69c) Flow Assurance in China West Crude Pipeline Transporting Multiple Waxy Crudes 14
Jinjun Zhang, Xiaolong Wang, Bo Yu, Hongying Li, Shanpeng Han

(74a) On Diffusion Coefficients, Relaxation Times of Oligonucleotides 17
Kal Renganathan Sharma

(74b) Thermodynamic Modelling of Br2-HBr-H2O Ternary System with Enrtl Model 18
Yue Yu, Chau-Chyun Chen

(74c) Evaluation of the Methods for Predicting Liquid Densities of High-Temperatures and High-Pressures Deep Petroleum Formations 19
Connor R. Albrecht, Akanni S. Lawal

(74d) Thermodynamic Model for Quaternary Aqueous Ca2+, Na+, Cl-, SO42- System 20
Sheik Tanveer, Chau-Chyun Chen

(74e) Understanding the Effect of Inhibitors on Controlling Asphaltene Precipitation and Aggregation Phenomena 21
Mohammad Tavakkoli, Miguel Garcia, Francisco Vargas, Ariana Melendez, Caitlin Garcia, Sarah Correa, Quinton Cox

(74f) PVT Modeling of Crude Oils: Development of Advanced Characterization Methods and Accurate Prediction of the Onset of Asphaltene Precipitation 22
Mohan Boggara, Miguel Garcia, Caleb Sisco, Francisco Vargas

(74g) Hydrate Plug Location during Depressurization Remediation Technique 23
Matthew W. Gilmer, Shirley Ike

(74h) Microfluidic Systems That Mimic Reservoir Conditions for Visualizing Asphaltene Aggregation and Deposition Near Wellbore Conditions (Electronic Poster Presentation) 24
Peng He, Yu-Jiun Lin, Francisco Vargas, Sibani Lisa Biswal

(74i) Flowloop Measurements of Hydrate Blockage Risk in â€˜Under-Inhibited Gas-Dominant Systems 25
Mauricio Di Lorenzo, Zachary M. Aman, Karen Kozielski, Bruce W. E. Norris, Michael L. Johns, Eric F. May

(74j) A New Quantitative Approach to Assess the Risk of Hydrate Blockage Formation in Subsea Flowlines 26
Bruce W. E. Norris, Zachary M. Aman, Michael L. Johns, Eric F. May

(74k) Prediction of Asphaltenes Solubility in Organic Solvents Via NRTL-SAC and COSMO-SAC 27
M. R. Islam, Meng Wang, Chau-Chyun Chen

(74l) Design for Mechanical Integrity in Thermal Mixing Points 28
Santhosh Shankar, Gonghyun Jung

(74m) Stimuli-Responsive Rheoreversible Fracturing Fluids for Unconventional Oil and Gas Extraction 29
Carlos A. Fernandez, Alain Bonneville, Hongbo Shao, Senthil Kabilan, David J. Heldebrant, David Hoyt, Tamas Varga, Lirong Zhong, Wooyong Um, Kenneth Carroll
(74n) Mitigating Downhole Flow Assurance Issues Using Specialist Modeling Software 30
Fabin Xu, Chunmin Zeng, Chong Zhang, Zhao Dong, Xiaohong Zhang, Richard Szczepanski, Adrian Ferramosca
(74o) Interaction of Oil and Gas Shale with Drilling Muds in Fracturing Fluids 31
Samyukta Koteeswaran, Peter Clark
(74p) Evaluation of the Methods for Predicting Dynamic Viscosity of High-Temperatures and High-Pressures Deep Petroleum Formations 37
Connor R. Albrecht, Akanni S. Lawal
(74q) The Van Der Waals Ten Commandments for Cubic Equations-of-State (Electronic Poster Presentation) 38
Akanni S. Lawal
(74r) Characterization of Critical Properties and Acentric Factors for High-Molecular-Weight-Fluids, Petroleum Fractions, Coal and Natural Products 42
Akindele Tododo, Neeraj Kumar, Akanni S. Lawal
(74s) Molecular Interpretation of Parameters in the Van der Waals Theory of Cubic Equations-of-State 46
Akanni S. Lawal
(74t) Rationale for the Parameterization of Van Der Waals Equations-of-State: The Lawal-Lake-Silberberg Equation As Classic Prototype 48
Akanni S. Lawal
(74u) Application of Artificial Neural Networks to Modelling Vapour–Liquid Equilibrium in the Fischer–Tropsch Reaction 49
Cornelius Mduduzi Masuku
(74v) Role of Surface Properties in Predicting Viscosity of Concentrated Coal-Water Slurries 50
Amrita Mukherjee, Sarma Pisupati
(74w) Optimizing Crude Oil Separations – a Study of Turbulence and Chemical Composition Influence on the Drop Size Distribution of Crude Oil in Maritime Conditions (Electronic Poster Presentation) 63
Stefan Ulrich, Sebastian Maass, Matthias Kraume
(74x) Prediction of High-Temperature and High-Pressure Dynamic Viscosity of Hydrocarbons By Van Der Waals Transport Equation of State 64
Kaijuan Zou, Akanni S. Lawal
(74y) Inversion of the Van Der Waals Parameters for Estimation of Binary Interaction Constants 67
Aldo Ortiz, Manish K. Lal, Akanni S. Lawal
(94a) Gas Evolution Rates in Crude Oil for Upstream Applications: A Literature Review 68
Alden Daniel, Clint P. Aichele, Sayeed Mohammad, James R. Whiteley, Gene Kouba, David Lavenson, Hariprasad J. Subramani
(94b) Fluid-Structure Interaction Modeling of a Subsea Jumper Pipe 69
Mohammad A. Elyyan, Yeong Yan Perng, Ali Najafi
(94c) A Novel Simple Approach to Describe Sloshing in FPSO Vessel Using CFD Simulation 79
Md. A Mahmud, Rafiqul Khan, Qiang Xu
(94d) Attributes and Behaviours of Crude Oils That Naturally Inhibit Hydrate Plug Formation 80
Brendan F. Graham, Zachary M. Aman, Agnes Haber, Bill Syddall, Paul Pickering, Michael L. Johns, Eric F. May
(114a) Reducing the Adhesion of Hydrates Using Durable Hydrate-Phobic CVD Polymers 81
Hossein Sojoudi, Karen Gleason, Gareth H. McKinley, Matthew Walsh
(114b) Avoiding Gas Hydrate Problems via Environmentally Friendly Gas Hydrate Inhibitors 82
Mert Atilhan, Santiago Aparicio, Bahman Tohidi
(114c) A New Application of Differential Scanning Calorimetry to Identify the Adsorption of Surfactants on Gas Hydrate Interfaces 83
Zachary M. Aman, Agnes Haber, Sarah J. Vogt, Michael L. Johns, Eric F. May
(114d) Formation of Hydrate Plug in Water-Decane-Natural Gas Systems with Varying Watercut and Inhibitor Concentrations 84
Yutaek Seo, Young Hoon Sohn, Jakyung Kim, Zachary Aman, Eric F May
(125a) Probing Asphaltene Deposition Using Microfluidic Devices 85
Yu-Jiun Lin, Mohammad Tavakkoli, Peng He, Sun Young Ji, Nevin Thunduvila Mathew, Yap Yit Fatt, John Chai, Afshin Goharzadeh, Francisco Vargas, Sibani Lisa Biswal
(125b) Evaluating Effect of Compositional Gradients on Asphaltene Stability in Reservoirs with CPA and PC-SAFT 86
Xiaohong Zhang, Merih Mustafa, Nuno Pedrosa, Richard Szczepanski

(125c) Cohesive Force Measurement Between Precipitated Asphaltene Solids 87
Shane A. Morrissy, Brendan F. Graham, Zachary M. Aman, Eric F. May, Michael L. Johns
(125d) Interfacial Properties of Asphaltenes at the Heptol-Brine Interface 88
Chuntian Hu, Ryan L. Hartman, Andrew Yen
(165a) Effect of Shear History on Restart of Gelled Waxy Crude Oil Pipeline 89
Youquan Bao, Jinjun Zhang
(165b) Evaluating Wax Gelation Risk for Restart Operations - Closing the Gap between Academic Research and Industrial Applications 92
Zhenyu Huang
(209a) Mining Big Data from Product & Process Simulations to Improve Designs 93
Alex Read, Domenic Barsotti, Angelo Flemings
(209b) Building Sustainable Data Quality Programs for E&p : Best Practices 94
Venkata Akella
(209c) A Distributed Modeling Framework for Monitoring Big Data in Plant-Wide Processes 95
Lester Lik Teck Chan, Zhiqiang Ge, Junghui Chen
Author Index

409505 Chemical Engineering: Is This the Golden Age?

Abstract:

Schlanger will offer his insights on the current opportunities in the chemical process industries in his talk entitled "Chemical Engineering: Is This the Golden Age?" on Monday morning, April 27. At the conclusion of his lecture, Schlanger will receive the AGILE Award in recognition of his leadership at CEVA Group, a $9 billion logistics supplier to industry worldwide; LyondellBasell, a major player in the plastics, chemicals, and refining industries; Arco Chemical; and at the other companies where he has held leadership positions.

399555 Adaption of Blosom (Blowout and Spill Occurrence Model) to High Performance Computing

Abstract:

The Blowout and Spill Occurrence Model (BLOSOM) is a multi-component computer program that simulates deepwater and ultra-deepwater oil blowouts from the oil well to the final fate or degradation point. The oil blowout trajectory includes a Lagrangian jet plume, underwater gas and hydrate kinetics, and weathering and degradation processes such as spreading, dispersion, evaporation, etc. While BLOSOM includes many processes, the different oil entities depend primarily on the environment and not on any other oil objects, which makes BLOSOM a prospect for high performance and parallel simulation. Minimizing the computing time for an oil spill's predicted trajectory is necessary for optimal spill response preparedness. BLOSOM currently incorporates C++ with such tools as the Boost Threading library and the OPENMPI message passaging interface (MPI) to parallelize the code which makes BLOSOM suitable for supercomputing simulation. In addition, further efficiency is obtained through neighbor tables and advanced search schemes of Geo-referenced raster formats. Results will include detailed descriptions of increased efficiency schemes, graphs of the oil trajectory over time, and a comparison of computing time simulations.

400677 Mechanism of Sloshing Minimization in Offshore Separator By CFD Simulation

Abstract:

Liquid sloshing in vessels caused by forced acceleration has been the subject of intense investigations for last several decades both by experiments and numerical simulations. Many studies are done to minimize the sloshing induced forces on the vessel internals and some studies focused on different ways to describe the sloshing patterns. In the literature, the focus of sloshing suppression has been varying different internals such as number of baffles, spacing, height and shape of these internals. The main factors behind sloshing reduction are alteration of frequency and energy dissipation. Few previous works have highlighted these reasons. However, there is a dearth of work in quantifying the frequency alteration and energy dissipation along with different combinations of vessel internals especially in 3D cylindrical tank. This present work tries to fill this gap by exploring sloshing minimization through Computation Fluid Dynamics (CFD) simulation for offshore separator. Several parameters were varied including baffles number, spacing, thickness, percentage opening subjected to wave induced vessel motions, motion frequency, amplitudes in various sea state conditions (e.g. surge, sway, roll, pitch). Parameters such as hydrodynamic force, pressure, moments, turbulent kinetic energy, height of the free surface, velocity, kinetic energy, sloshing frequency etc are used to quantify the sloshing intensity. In addition, visual inspections of sloshing motion are done through gas-liquid/oil-water interface fluctuation, streamlines, vector profiles. Correlations between these parameters influencing the sloshing intensity will be developed to understand the sloshing mechanism.

399942 Detailed Flow Modeling of Erosion in Subsea Equipment

Abstract:

Particulate erosion is one of the common flow assurance problems in subsea and other upstream equipment. Erosion causes failure of equipment and is a safety hazard. A good understanding of the causes of erosion is essential before ways of mitigating them may be designed and verified. Erosion is caused by the interaction of fluid and particle flow with the walls of the equipment. The fundamental basis of erosion is the fluid mechanics of particulate flow in a fluid medium. Computational fluid dynamics has proven to be useful in understanding this multiphase flow and erosion. In this paper we demonstrate the importance of using appropriate boundary treatment for particle interactions with walls for a realistic prediction of particulate erosion.

Inclusion of microscopic surface roughness of walls affects post collision rebound of the particles and then it has an important influence on the areas of the equipment prone to erosion. While validating simulation results with experiments done by Mason et. al. it was found that ignoring surface roughness, incorrectly predicts the amount and extent of erosion. The other effect that is important to include is the change in the fluid domain due to equipment material removed by erosion. Change in fluid domain changes the particle fluid dynamics and thus impacts the new areas of equipment where erosion might start.

We present a CFD implementation of the above said effects in ANSYS CFX and ANSYS Fluent and its validation with the experimental data of Mason et. al. We then showcase the effects of this implementation on typical equipment found in subsea fields

400015 Erosion Model Development for High Concentration Slurry Flow Applications

Abstract:

Computational fluid dynamics (CFD) combined with the discrete element method (DEM) has been used to solve solid particle transport problems for over two decades [1], but only recently the development has been shifted to focus more on liquid slurry flows [2] due to the scarcity of published data and the difficulty in handling high concentration solid flows both experimentally and numerically. This talk will present the development of erosion models using the CFD-DEM approach because of many upstream flow assurance needs, especially in the oil sand operation and hydraulic fracturing.

In contrast to low concentration solid flows, the particle interactions cannot be ignored under high solid concentrations. The commonly used DPM model that ignores such interactions can produce unphysical results showing particle concentration exceeding the packing limit and overestimate the erosion rate by a large degree. However, the hybrid CFD-DEM approach can be expensive and tricky due to its computational cost that limits the number of tracked particles. In addition, models for particle/fluid interactions must be used and that adds another layer of uncertainty. DEM method is preferred due to its wide use in gas-solid flows, but its application in liquid slurry flows has been limited. The Eulerian-Granular model based on the kinetics theory has been successfully applied to dense slurry flows recently [3], and it has been shown that it is capable of predicting solid concentration profiles and pressure drops satisfactorily, but it lacks a proper mechanism to capture erosion phenomenon which requires particle tracking methods that can mimic solid flows in high concentration.

Works will be presented that validate the hybrid CFD-DEM approach by correctly modeling the solid concentration profiles and pressure drops as captured by the Eulerian-Granular model. Furthermore, the erosion correlation that often used in the gas-solid flows was fine-tuned to reflect the impact of liquid flows.

References

1. Snider, D. M.: "An incompressible three-dimensional multiphase particle-in-cell model for dense phase flows," *J. Comp. Phys*., **170**, pp. 523-549, 2001.
2. Tsai, K., E. Fonseca, S. Degaleesan and E. Lake, "Advanced computational modeling of proppant settling in water fractures for shale gas production," SPE Journal **18,** pp. 50-56, 2012.
3. K. Ekambara, R. S. Sanders, K. Nandakumar and J. H. Masliyah, "Hydrodynamics simulation of horizontal slurry pipe flow using ANSYS-CFX," *Ind. Eng. Chem. Res*, **48**, pp. 8159-8171, 2009.

399554 State of Utah Energy Research Triangle (University of Utah, Brigham Young University, Utah State University) Project for Characterization of Uintah Basin Waxy Crude Oil Deposition in Pipelines

Abstract:

This presentation provides an overview of the subject project, and discusses opportunities for enhanced pipeline transport of Uintah Basin (Utah) waxy crude oils. The project involves characterization of the viscosity and wax solubility of Uintah Basin waxy crudes, with and without feasible diluents. Diluents and modifiers include regionally available crude, other raw or refined hydrocarbons, condensates, surfactants, and possibly water (as an external emulsion phase). Wax solubility and resultant crude oil viscosity are key parameters for deposition models specific to Uintah Basin derived waxy crudes. Modeling has been implemented to evaluate techniques for process control at the wellhead (or centralized facility) and at the refinery. Evaluations have been undertaken of the rate and spatial profile of deposition within appropriately-sized pipelines at pipeline-ambient temperature. This leads to an understanding of possible pigging regimes and determination whether diluents alone may be used to reverse pipeline deposition. Additional considerations of the study include process control, land use, and environmental aspects of pipelining Uintah Basin waxy crude oils.

The Effect of Computational Fluid Dynamics Methodology on Solid Particle Erosion Predictions

Deval Pandya[1,2], Ronnie Russell[3] and Brian Dennis[1], (1)The University of Texas at Arlington, Arlington, TX, (2)Shell Global Solutions, Houston, TX, (3)Baker Hughes, Houston, TX

Extended Abstract

Erosion is a complex phenomenon because many factors affect erosion in oil and gas applications. Tool geometry and material, particle size, concentration and shape, fluid velocity and viscosity are some of the factors that affect the erosion rate and location. Oil and gas equipment manufacturers, operators, and service companies are all interested in methods that can predict erosion. A reliable erosion prediction method, can save money and lives by helping to predict equipment failure accurately and to maintain equipment.

A computational fluid dynamics (CFD) - based erosion model offers an inexpensive and rapid solution to the erosion modeling problem. A CFD-based erosion modeling procedure essentially consists of three steps: (1) Flow modeling using CFD, (2) Discrete phase modeling (DPM) to obtain particle trajectories and wall interactions, and (3) Erosion calculations based on empirical equations. Each step affects the accuracy of the final erosion prediction and is critical for overall analysis.

This work focuses mainly on the effect of a grid strategy on the accuracy of predicted erosion rates. Simulations are performed for various grid sizes to replicate experiments conducted for four different flow conditions considering two flow rates; 50 ft/sec and 85.8 ft/sec and for two particle sizes; 256 microns and 25 microns (Russell, 2004). The test was performed on a 90-degree elbow with a 1-in. inner diameter. Inconel 718, Nickel Alloy 825, 25% Cr, Nickel Alloy 925, and 13% Cr L-80 material elbows were placed in series. The sand concentration was approximately 1% of the liquid weight and 0.38% by volume (Russell, 2004). The erosion rate was measured at 230 different locations on the elbow. The experimental set up is shown in Figure 1, and the details of erosion test fixture (elbow) are presented in Figure 2. A proprietary erosion model was used to predicted erosion.

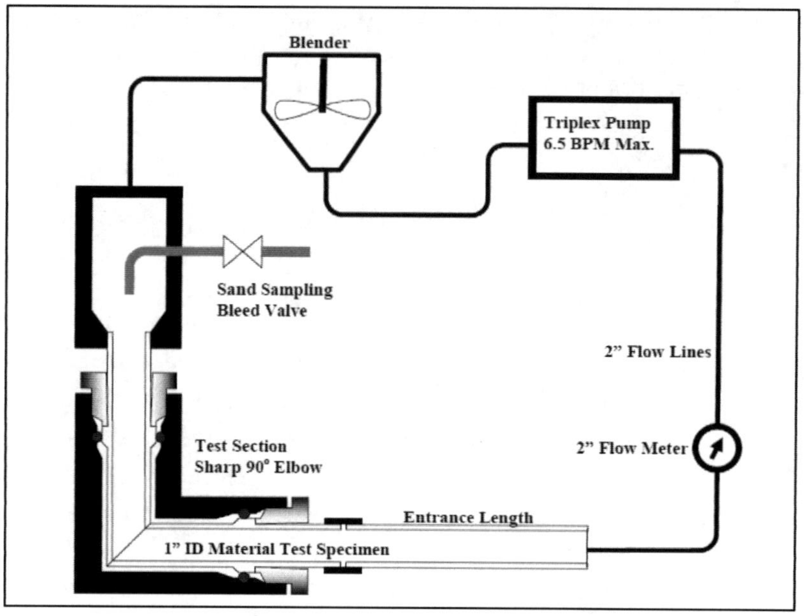

Figure 1 Experimental setup for the erosion experiment (Russell, 2004)

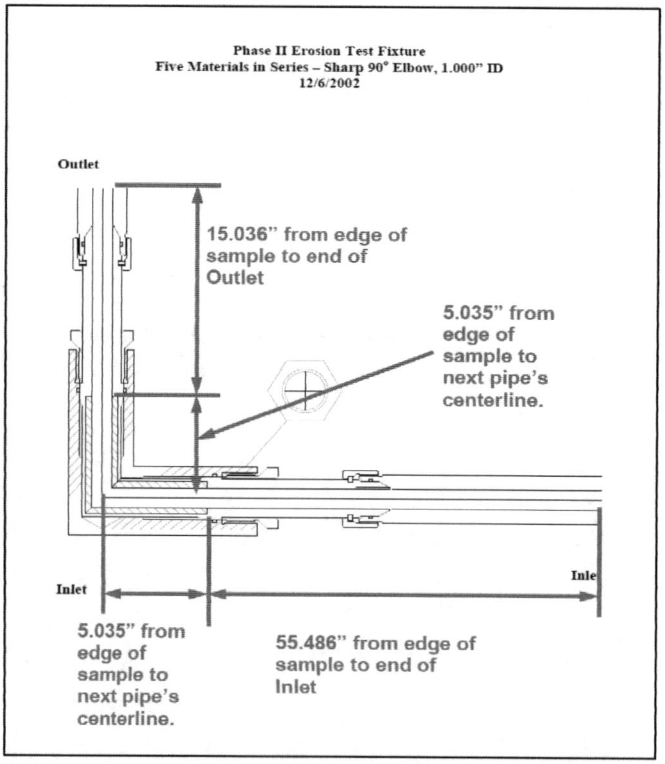

Figure 2 Erosion test fixture

The Lagrangian approach was applied for solid particle tracking after simulating the flow field. Preliminary grid studies were performed on a symmetric half model of a sharp elbow using more than 15 mesh sizes. The wall y+ varied from 1 to more than 2000, with the number of elements varying from 1 million to 17 thousand, respectively. These studies showed that erosion rates were extremely sensitive to mesh size. An elaborate study was performed on more than 20 grids for a full model. The results confirmed the high erosion rate sensitivity of the mesh, with the mean percentage error ranging from 35% to 120% in some cases. Each grid had a different wall y+ value for each case. Consequently, results in terms of y+ values are presented.

Figure 3 plots minimization functions defines as Fn= $\left\{1 - {CFD\ Erosion\ value}/{Experimental\ Erosion\ value}\right\}$ for three different y+ ranges for all four test cases.

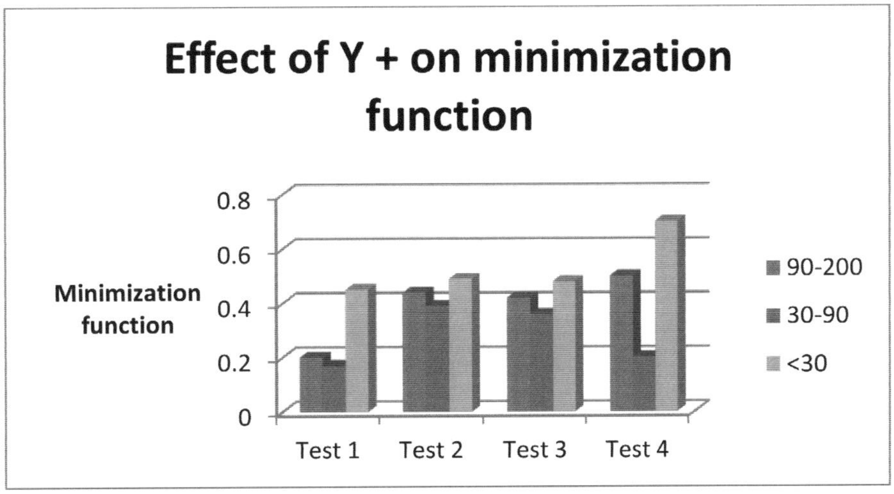

Figure 3 Minimizaiton function comparison (Pandya,2014)

To remove this dependency on grid size, new models based on y+ values were developed. These models showed great improvement in erosion rate predictions, irrespective of grid (i.e., by applying a y+ 90-200 model on a grid with that y+ range gave similar results to applying a y+ 30-90 model to grid with y+ 30-90).

Table 1 Goodness of parameter fit for mesh specific models

	Sum of squares Error	R -square	Root mean square Error	Percentage error
y+ 90-200	1.71E-07	0.992	3.70E-05	37
Y+ 30-90	5.70E-07	0.9957	2.50E-05	34
Y+ < 30	1.12E-07	0.9927	1.10E-05	31

Finally, a sensitivity study was performed to determine the effect of mesh size on DPM concentrations. For every test case, the DPM concentration showed a standard deviation of 10 to 20, compared to approximately 1 for other output parameters. These results indicate that the DPM parameter is highly sensitive to grid size. The effect of the near wall on solid particle concentration, comparing DPM (Lagrangian) model with eulerian granular model is presented. Future work to develop eulerian model-based erosion model is recommended.

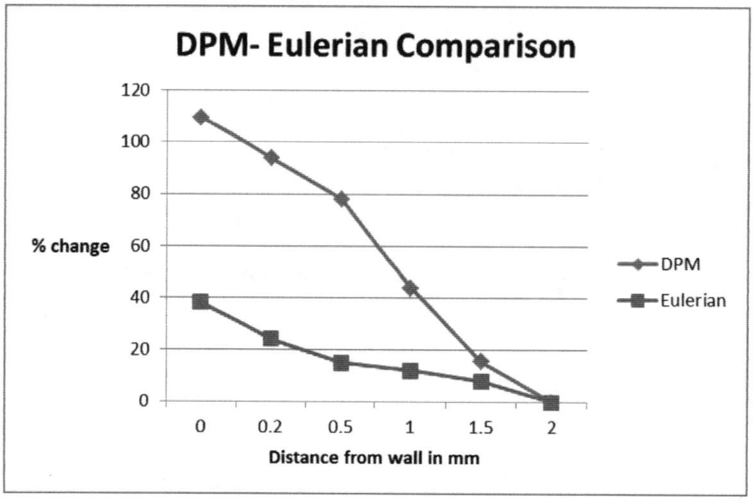

Figure 4 DPM- Eulerian model comparison for near wall solid particle concentration in a 90^0 elbow(Pandya,2014)

References

Russell, R., Shirazi, D., & Macrae, J. (2004). A new computational fluid dynamics model to predict flow profiles and erosion rates in downhole completion equipment. Paper presented at the *SPE Annual Technical Conference and Exhibition. SPE-90734-MS*

Pandya, D., Dennis, B. and Russell, R. (2014). An Improved Computational Fluid Dynamics (CFD) Model for Erosion Prediction in Oil and Gas Industry Applications. Presented at the ASME 2014 33rd International Conference on Ocean, Offshore and Arctic Engineering -San Francisco, California. http://dx.doi.org/10.1115/OMAE2014-23136

Russell, R. and Marsis, E. (2013). A State-of-the-Art Computational Fluid Dynamics Simulation for Erosion Rates Prediction in a Bottom Hole Electrical Submersible Pump. Presented at the SPE

Heavy Oil Conference-Canada. Calgary, Alberta, 11-13 June. SPE-165452-MS

http://dx.doi.org/10.2118/165452-MS

398512 Evaluating Wax Yielding with a New Model

Abstract:

Determination of the yield stress of waxy crude oils is important to understand the restart of pipelines containing oil that has gelled during a shutdown. While it has been observed by several researchers that the pressure required to restart flow is not directly proportional to the yield stress, this rheological value is nevertheless important in estimating an approximate value for the restart pressure.

We have worked internally and with different collaborators to experimentally and theoretically study the rheology of waxy oils. One of the outcomes is a model called the IKH (Isotropic Kinematic Hardening) Model recently developed by MIT (Dimitriou et al, J. Rheology, 57, 2013). This model considers the kinematic hardening process related to the plasticity of the gel and the isotropic hardening associated with the structural changes of the gel.

Our rheological experiments with a number of crude oils as well as model oil systems have shown that, upon application of a constant shear stress, the gel creeps until a critical strain value of about 1% is reached, whereupon a catastrophic failure is observed.

In this presentation, we discuss these experimental results and a way to model the results with the IKH model. Specifically, the ability of the model to predict the experimentally observed breakdown strain limit and the implications on predicting restart pressure in pipelines are discussed.

399041 A Self-Sufficient Wax Deposition Model for Horizontal Gas-Oil Slug Flow

Abstract:

A wax deposition model based on the laboratory flow loop experiments was developed, validated, and explained in detail in this study. The circumferential area is discretized into top and bottom parts to capture the non-axisymmetric of wax deposition behaviors. The deposit is assumed to be smooth and traps no gas inside. The physical unknown constants used in the model to capture the non-conventional deposition mechanism are introduced. The practical closure relationships of these unknown constants were developed by using Levenberg-Marquardt algorithm with 2-in flow loop deposit thickness and wax fraction data, where Garden Banks condensate was used as a testing fluid. This model with closure relationships serves as a self-sufficient predictive tool improved over previous self- and non-self-sufficient wax deposition prediction models.

401457 Flow Assurance in China West Crude Pipeline Transporting Multiple Waxy Crudes

Abstract:

The China West Crude Pipeline (CWCP), commissioned in 2007, starts from Urumqi, the capital of Xinjiang Uygur Autonomous Region, and ends at Lanzhou, the capital of Gansu Province. The capacity of the 1838 km, 813/711 mm O.D. pipeline is 20 million tons annually (about 400 kBPD). Connected with the Kazakhstan-China crude pipeline and three major oilfields in Xinjiang, supplying crudes for two refineries, this pipeline acts as a strategic channel in western China for petroleum transportation.

Among the four crudes transported through this pipeline, three are waxy crudes with the pour point possibly as high as 20 ^{o}C. Flow assurance has been a major challenge in waxy crude pipeline. For the CWCP, flow assurance has been much more complicated as a result of the considerably variable cold flow properties of the crudes (the standard deviation of the pour point may be as high as 5 ^{o}C), combined with cold environment (the soil temperature at the buried depth of the pipeline may drop to 2 ^{o}C or even below in winter), highly variable transporting requirements of multiple crudes, complex thermal interaction between the crude line and the product line which was buried 1.2 m apart from the crude line.

Actually, more than 80% of crude production in China is waxy crude, with the most typical being the crude produced in Daqing oilfield, the largest oilfield in China, which has a waxy content of 25 wt% and a pour point as high as 35 ^{o}C. Currently in China, pipeline companies operate 30,000 km of crude oil pipelines, with most of them transporting single of waxy crude or blend of waxy crudes. Waxy crude rheology and flow assurance of waxy crude pipelines has been a major field of research in China. Thanks to our experiences and achievements obtained in decades of researches in waxy crude rheology and flow assurance of crude oil pipeline, we successfully tackled these problems, ensuring safe and efficient operation of this pipeline. The main technological advances are as follows.

1. Batch Transportation of Multiple Crudes Treated with PPD

Pour point depressant (PPD) modification has been successfully and widely used in China's waxy crude pipelines since the mid-1990's. The PPD-modification was promoted to the CWCP that transports multiple waxy crudes in batch, that is, to carry out proper PPD treatment according to the cold flow properties and the transportation task of the crudes. Experiments were firstly conducted for determination of the PPD dosage and treatment temperature, and a dosage of 20-50ppm and a treatment temperature of 55 °C were found proper for the crudes to be transported. More importantly, for transporting PPD-treated waxy crude through long distance pipeline, the accurate understanding of the shear and thermal effect is crucial, since the high shear imposed while flowing through pumps and low but long-term pipe flow shear may make the reduced pour point and viscosity recover more or less, and improper reheating to the treated crude may do harm to the flow improvement of the PPD. These phenomena are called the shear effect and thermal effect, respectively. We accurately obtained this shear and thermal effect by quantitatively simulating pipelining process by using the approach we originally developed that uses the entropy generation of viscous flow as the rule of similarity between experimental simulation and the industrial pipeline flow. The use of PPD modification successfully enhanced the flow assurance capability of the pipeline in winter operation. As a result, fuel consumption of this pipeline may be reduced by 83% compared with operation via heating the crudes.

2. Intermittent Operation

Because of the need for full delivery of the Tuha crude at the intermediate station at Yumen, the downstream segment of the CWCP, 792 km long, is operated normally in an intermittent state, with averagely shut down for 1 day after 5.4 days of running and 25.4 h of mean time for shutdown. The intermittent operation of waxy crude pipeline challenges flow assurance because its possibility of pipeline restart failure could be much higher than that of the continuously operated pipelines. This is mainly attributed to the less thermal storage in the soil and the resultant quicker cooling of the oil. To ensure safe operation of the intermittently operated pipeline, specifically to ensure restartability of the shutdown pipeline, it is crucial to fully understand the thermal and hydraulic characteristics of the pipeline. However, since the endlessly and irregularly varying soil temperature field which is a function of the pipeline operation history, it is very difficult to obtain accurate simulation result. To tackle problems from the intermittent operation of the CWCP, we used the PPD to improve the intrinsic safety of flow on one side, and on the other side we developed a model and corresponding algorithm for numerical simulation of the intermittently operated pipeline. Based on the understanding of the thermal and hydraulic characteristics thus obtained, and the assessment of restartability using the reliability-based limit state approach mentioned below, scenarios for intermittent operation were further optimized. Field tests were conducted to validate the reliability of this operation, and finally intermittent operation became part of the normal operations of the CWCP. Intermittent operation was also carried out in the whole pipeline from January to April 2009, when the CWCP was impacted by the global financial crisis and the throughput was reduced to one fourth of the designed value.

3. Assessment of the Restartability of Waxy Crude Pipeline with the Probabilistic Approach

Pipeline restart failure as a result of the gelation of waxy crude has been a major threat of the safe operation of waxy crude pipeline. Many influencing factors such as crude properties, operation and environmental parameters have the attribute of uncertainty. Current restart calculation methods take no consideration of this uncertainty, including the unsteady heat transfer and flow algorithm as well as the traditional force balance equation. The reliability-based limit state approach is an effective method to tackle problems with uncertainty, and this probabilistic approach has been successfully used in some fields such as aerospace industry and structural engineering. In the pipeline industry, ISO issued the first standard of reliability-based limit state method for the design and assessment of oil and gas pipeline structure (ISO 16708) in 2006. Inspired by this idea, we, for the first time, introduced this approach to assess the restartability of waxy crude pipeline so that the influence of parameter uncertainty can be taken into consideration, and the restartability can be assessed quantitatively in the way of restart failure probability. Consequently, the lowest allowable pumping temperature and the maximum allowable time for pipeline shutdown can be determined. The key problems in development of this approach include, firstly, the limit state equation that can accurately describe the complex unsteady heat transfer and unsteady flow of non-Newtonian fluid, and secondly, an efficient algorithm for reliability analysis because of the extremely heavy load of probability computation resulted from the thousands of random sampling of the stochastic variables and the consequent thousands of the numerical solution of the differential equations of unsteady flow and heat transfer.

4. Transporting Hot and Cold Crudes in Batch.

Nearly 50% crudes transported through the CWCP can be pumped without heating or any other treatment such as PPD modification in the whole year, and the remaining crudes have to be heated to 55 °C at the initial station in winter to achieve PPD-modification. Heat consumption may be greatly reduced if the crudes having good flowability are pumped without heating. However, this way of transporting hot and cold crudes in batch is difficult and complex because the crude temperature, pressure, and soil temperature are always in the complex alternating process, and the hydraulic and thermal conditions interact with each other. Therefore, flow assurance is much more complicated than the normal hot crude pipelines. Moreover, the structural safety of pipeline under the alternating load is also a critical concern. Via numerical simulations on this transportation process using software we developed, we well understood the hydraulic and thermal characteristics including normal and restart operation, and further optimized the pumping scenarios. The structure reliability under the conditions of batching hot and cold crudes was confirmed after the analysis on the strength, stability and fatigue life of the pipeline structure. The validity of this pumping scenario was tested in field trials. It has become one of the regular operations in the CWCP. Each year in the early winter from November and December and the early spring from March and April, the Tuha crude is transported via PPD-treatment, while the remaining crudes are transported without heating.

According to various transportation requirements and by the integrated use of these technologies, the CWCP has successfully risen to a variety of transportation challenges and well met the demands either from the oil fields or petrochemical enterprises. The pipeline has been operated safely, efficiently and flexibly since 2007.

398087 On Diffusion Coefficients, Relaxation Times of Oligonucleotides

Abstract:

DNA microarrays have been applied to gene expression, genotyping and sequencing. The hybridization step between probe and target molecules may take several hours. The understanding of the hybridization kinetics is important. Some investigators have found the hybridization overall reaction rate to be reaction rate limited in the 90s and some other investigators in the early 00s have found the overall hybridization reaction rate to be diffusion limited (Pascault and Zhou, 2008). They go further to improve the overall hybridization reaction rate by providing convective mass transfer to decrease the diffusion limitations. They obtain these recommendations after performing computer simulations by finite element numerical solutions using Comsol Multiphysics software package. The length of the DNA target molecules range between 50 – 350 bp. The probe molecules are shorter than the target molecules and have about 10-60 bp of Adenine, Cytosine, Guanine and Thymine. The surface concentration of probe molecules range between 1.2 x 10^{-10} – 4 x 10^{13} probes/cm^2 or 1 attomoles/m^2 to 0.1 micromoles/m^2. The diffusion coefficient values for the base pairs decreased with the length of the oligonucleotide. The variation of diffusivity values as a function of the length of the oligonucleotide from different investigators is shown in Figure 1.0. With a R^2 value of 0.9956 the following curve could be seen from the trendline except for the data point from Ferrari and Bloomfied given in Pascault et. Al. [2008]. Their value is also given as a red rectangle in Figure 4.3. The diffusion coefficient can be written as follows;

$$D_{pl} = al^{0.639}$$

The relaxation time for oligonucleotides can be written as;

$$\tau_r = b(l^{1.639})/RT$$

Where l is the lengt of the oligonucleotide in terms of number of base pairs, and D_{pt} is the binary diffusivity of the probe molecules in target molecules. A_1 was found to be 6 x 10^{-10} m^2/s from the regression tool in MS Excel 2010 for Windows using a HP Compaq Elite 8300 desktop computer with i7 intel processor. R the universal molar gas constant, T the absolute temperature. Target molecules of 1 – 1000 pM were considered. The channel height of the microfluidic elements range was 50 – 100s μm. Target microarray spot size considered was 200 μm. The two dimensional microfluidic chamber had a height of 100 μm.

Figure 1.0 Diffusion Coefficient of Oligonucleotide as a Function of Size

399668 Thermodynamic Modelling of Br2-HBr-H2O Ternary System with Enrtl Model

Abstract:

The storage of energy has always been researchers' concern for decades. As an improvement of fuel cells, the technique of redox flow cells provides us a great balance between energy density and electric-to-electric efficiency. The H_2-Br_2 flow cell system has the advantages in reaction kinetics, power density, energy storage capacity, simplicity, and low cost on materials. In the system, there are molecular components including H_2, Br_2, HBr, H_2O, and ionic components such as H_3O^+, Br^-, even H_5O_2+. Proper thermodynamic modelling of the complex electrolyte solution is of great importance for further development of the H_2-Br_2 battery system. Wozny looked into phase equilibria of HBr/H_2O system, and a correlation for only pressure was put forward in 1981. Applying electrolyte non-random two liquid model, this work investigates the adjustable pair parameters between molecular-molecular components, molecular-ionic components, and ionic-ionic components. Experimental data of phase equilibria, solubility, and molar enthalpy are used to regress these parameters for both HBr-H_2O binary system and Br_2-HBr-H_2O ternary system. It turns out that ENRTL model works in excellent precision for this system. The redox potential of flows in the cell system can also be evaluated with this model.

400429 Evaluation of the Methods for Predicting Liquid Densities of High-Temperatures and High-Pressures Deep Petroleum Formations

Abstract:

Six currently used methods for predicting liquid densities of high-temperature and high-pressure (HTHP) deep-water petroleum formations are evaluated for ease of implementation and applications for flow assurance modeling, pipeline flow simulation and reservoir simulation of carbon dioxide sequestration in the depleted oil-reservoirs. The evaluated techniques include the various modified versions of the *Peng-Robinson EoS*, the *high-temperature, high-pressure, volume-translated Peng-Robinson EoS*, the *volume-translated Soave-Redlich-Kwong EoS*, the *generic Van der Waals cubic EoS*, the *modified Tait equation* and the *perturbed-chain statistical associating fluid theory EoS*. The HTHP liquid density data used in the evaluation of the methods for HTHP liquid densities are reported in the literature by Enick Group at U of Pitt, Caudwell-Trusler-Wakeham Group at Imperial College London, Ducoulombier et al. in France, the Doolittle specific volumes of n-Alkanes (nC_7 to $C_{40}H_{82}$) and the Sage-Lacey volumetric data of CO_2, N_2, H_2S, C_1 to nC_{10}, including the isomers of iC_4 and iC5. While all the evaluated methods gave various degrees of uncertainties at HTHP conditions (due to the judicious choice of the parameters required to implement PC-SAFT and the volume-translated techniques), the VDW generic equation of state (with covolume temperature function) performs better and in overall agreement with the measured liquid densities at the HTHP conditions of deep-water petroleum formations.

The analysis of model results highlight the capability of the various techniques to account for the liquid densities measured for some isomers investigated at HTHP conditions as well as for high-molecular-weight fluids. The HTHP VT-PREoS provides minor improvements over the original 1976 design of PREoS, but neither EoS is capable of accounting for the effect of isomer structural differences on the measured liquid densities at HTHP conditions and significant deviations are observed at HTHP liquid densities of heavier molecular-weight hydrocarbons. However, the PC-SAFT EoS, with pure component parameters derived from the literature provides modest improved liquid density predictions over those obtained with the PREoS or HTHP VT-PREoS or volume-translated SRK EoS.

Overall, the VDW generic cubic EoS (with embedded covolume temperature function) provides more accurate predictions of liquid densities at HTHP conditions associated with the deep-water petroleum formations than all the methods of predicting liquid densities that are analyzed in this investigation. Even though the PC-SAFT EoS shows improved predictions of liquid densities over the Tait correlations and over those EoS (without the embedded covolume temperature function), it still over predicts the experimental liquid densities at HTHP conditions. The VDW generic cubic EoS is recommended for all analyses of flow assurance modeling. But, we are still performing similar analysis for derivative properties (isothermal compressibility, volume expansivity, bulk modulus, isobaric and isochoric heat capacities, Joule-Thomson coefficient and speed of sound) at HTHP conditions associated with the deep-water petroleum formations.

Keywords: cubic equation, Tait equation, liquid density, covolume temperature, heavy hydrocarbon

Thermodynamic Model for Quaternary Aqueous Ca^{2+}, Na^{+}, K^{+}, Cl^{-} System

Abstract:

Symmetric Electrolyte Non-Random Two Liquid model (eNRTL) is selected as the thermodynamic framework to model the quaternary aqueous Ca^{2+}, Na^{+}, K^{+}, Cl^{-} electrolyte system. This is a part of a large effort to develop a comprehensive thermodynamic model for high salinity produced water in oil and gas production. Each binary system (electrolyte-water or electrolyte-electrolyte) needs three pairs of binary parameters which are directly correlated with excess Gibbs energy, excess enthalpy and excess heat capacity of the system concerned. These binary parameters are regressed with thermodynamic and phase equilibrium data including salt solubility data available in literature. Binary parameters are obtained for $CaCl_2$-water, $CaCl_2$-KCl and $CaCl_2$-NaCl binary systems. These binary parameters are integrated with the rest of the parameters readily available in literature. The model shows excellent agreement with literature data with temperature from 0°C - 200°C and electrolyte concentration up to saturation. This model is being extended to cover Mg^{2+}, Ba^{2+}, Sr^{2+}, SO^{4-} and $HCO3^{-}$.

399251 Understanding the Effect of Inhibitors on Controlling Asphaltene Precipitation and Aggregation Phenomena

Abstract:

Understanding the Effect of Inhibitors on Controlling Asphaltene Precipitation and Aggregation Phenomena

Mohammad Tavakkoli[1, 2], Miguel Garcia-Bermudes[1], Ariana Melendez[1], Caitlin Garcia[1], Sarah Correa[1], Quinton Cox[1] and Francisco Vargas[1, *]

1. Department of Chemical and Biomolecular Engineering, Rice University, Houston, Texas, USA.

2. Department of Chemical and Petroleum Engineering, Sharif University of Technology, Tehran, Iran.

* fvargas@rice.edu

Asphaltenes tend to precipitate and deposit during oil production and clog the wellbore and production facilities. This can cause an excessive downtime and high cleaning costs running into several millions of dollars. Therefore, a mitigation technique such as using asphaltene inhibitors is necessary. To improve the efficiency of the current inhibitors or develop a new generation of asphaltene inhibitors, it is necessary to study the effect of inhibitors on controlling asphaltene precipitation, aggregation and deposition processes.

In this study, we have investigated the effect of different commercial inhibitors on asphaltene precipitation and aggregation processes. It needs to be clear whether the inhibitors make changes on the asphaltene precipitation such as changes on the onset and the amount of precipitation or they only affect the asphaltene aggregation process by dispersing asphaltene aggregates and changing the particle size distribution. To do so, three commercial inhibitors were used and asphaltene precipitation and aggregation phenomena were studied using two different techniques: "Direct spectroscopy" and "indirect method". The experiments were conducted at ambient conditions and asphaltenes were precipitated using light normal alkanes such as n-pentane and n-heptane. Using direct spectroscopy technique, we found that asphaltene particle size distribution changes by adding the inhibitors to the crude oil especially at high amount of the added precipitant. The inhibitors break the asphaltene agglomerates and form smaller size aggregates. Indirect method, which is a combination of gravimetric and spectroscopy methods and is a novel technique for studying asphaltene precipitation, showed that there are no changes on the onset and the amount of asphaltene precipitation when the inhibitor is added to the crude oil prior to adding the precipitant. Therefore, one can conclude that the commercial inhibitors can disperse the already precipitated and aggregated asphaltenes and form smaller particles, but it is still under investigation whether this working mechanism of inhibitors always prevents asphaltene deposition in the wellbore or may worsen the problem in some cases.

Keywords: Asphaltene Precipitation, Asphaltene Aggregation, Commercial Inhibitors, Indirect Method.

400030 PVT Modeling of Crude Oils: Development of Advanced Characterization Methods and Accurate Prediction of the Onset of Asphaltene Precipitation

Abstract:

With ever-increasing complexities in production environments – including gas injection for EOR or deep sea Pressure-Temperature conditions – and the high cost of sampling reservoir fluids, accurate modeling of PVT properties with minimal field data is critical. These difficulties necessitate a continual reevaluation of the characterization methods for crude oils as well as the Equation of State (EoS) models used to predict PVT properties. The impact of such accurate PVT characterization and modeling of crude oils in reservoir simulations and EOR applications cannot be over-emphasized. In this work, we review well known methods for characterizing the plus fraction of black oil and gas condensate mixtures and compare various equations of state in predicting the phase behavior of the reservoir fluids in a wide range of pressures and temperatures. The equations of state studied in this work are the conventional cubics (i.e. Soave-Redlich-Kwong and Peng-Robinson) with and without volume correction and PC-SAFT. We will revisit previous characterization methods and discuss the development of new, simplified characterization methods. A thorough discussion is presented on the characterization procedures, advantages and disadvantages of the different methods, and the performance in the prediction of PVT properties.

From the flow assurance stand point, accurate prediction of asphaltene precipitation is absolutely critical for (a) understanding the likelihood of asphaltene deposition problems, (b) implementing proper mitigation strategies and (c) developing suitable EOR strategies, as gas injection can potentially increase the propensity of asphaltene precipitation. In this regard, we build on our previous work in the use of PC-SAFT EoS along with simplified correlations relating the PC-SAFT parameters of Asphaltenes, Aromatics and Resins to their MW (or a MW distribution) and a new parameter called aromaticity. Further, the accuracy of newly developed characterization methods in the prediction of bubble pressure curves and asphaltene onset curves will be discussed.

401699 Hydrate Plug Location during Depressurization Remediation Technique

Abstract:

During the remediation of hydrate blockage in a pipeline, certain key parameters such as the location of the first plug, the velocity of the plug upon release, the gas volume fraction upstream and downstream of the pig, etc. are important to know for safe handling of the process. Typical remediation of a blocked pipeline would include alternating cycles of depressurization and inhibitor injection to safely handle plug movement.

Utilizing operating information during the remediation process, MSi Kenny has developed a toolkit that estimates the hydrate plug location based on alternating between inhibitor injection and depressurization method of remediation. The purpose of this model is to determine the plug location and calculated downstream gas volume fraction.

The model uses the pressure rise in the pipeline during the inhibitor injection cycle (with known methanol injected volume), the subsequent pressure drop and liquid offtake volume during the depressurization step. Using these inputs the model iteratively calculates the first mathematically stable plug location and downstream gas volume fraction. These calculations are currently based on basic equations of state. Initial comparison of the model to multiphase flow simulations show the results to be in close agreement.

This model offers a tool to predict potential hydrate plug locations for this particular remediation technique. With an estimate of the hydrate plug location, operators can have a better idea of hydrate plug movement risks, more specific hydrate remediation locations, or simply a secondary method to verify plug locations.

400322 Microfluidic Systems That Mimic Reservoir Conditions for Visualizing Asphaltene Aggregation and Deposition Near Wellbore Conditions (Electronic Poster Presentation)

Abstract:

Asphaltene deposition is well known as a cause of production rate decline and various other operational problems in most oil production processes. Asphaltenes are typically stable under virgin reservoir condition, however may be destabilized and precipitate from crude oil and deposit on the internal surface of equipment or near-wellbore reservoir during production and transportation. Asphaltenes precipitation and deposition are often due to changes in thermodynamic properties, and pressure has a large effect on the flocculation of asphaltenes. Previous studies have investigated these phenomena in bulk volume (e.g. in stainless steel pipe) but visualization are often missed in these "black-box" type experiments, and the results from such macroscopic observations could not reflect the nature of asphaltenes at micro-scale. Micro-scale behaviors of asphaltenes are important for the fundamental understanding of origins of asphaltene flow assurance problems. Prevention and mitigation of asphaltene problems may also need to be applied at the source of crude oil in wellbore conditions and in porous media. Hence there is a great need to fill in the gap by visualizing asphaltene aggregation and deposition behaviors at micro-scale near wellbore conditions. In this report, we will demonstrate preliminary results of such experiment using microfluidic systems. The variation of asphaltene solubility in model oil with change of pressure will be visualized in micromodel, and flow experiment of asphaltene deposition on porous media surface will also be demonstrated. These microfluidic experiments has the potential to provide new information for fundamental research at wellbore conditions as well as a mean of fast inhibitor development.

400338 Flowloop Measurements of Hydrate Blockage Risk in â€'Under-Inhibited Gas-Dominant Systems

Abstract:

In conventional flowlines, gas hydrate formation is mitigated through the injection of thermodynamic hydrate inhibitors (THIs) that shift the hydrate stability criteria toward high pressure and lower temperature. Complete hydrate avoidance with THIs substantially increases infrastructure and operating costs, which may affect the economic viability of deepwater assets. Reducing the rate of THI injection below the complete inhibition threshold may translate to significant cost savings, but adoption of this practice requires a rigorous experimental confirmation that such under-inhibition will not increase the risk of hydrate blockage formation. In this study, we have deployed a high-pressure sapphire autoclave and high-pressure single-pass flowloop to measure hydrate formation and resistance-to-flow (motor torque or pressure drop) in a matrix of under-inhibited conditions. 20 autoclave studies were performed with deionised water containing 0-20 wt% monoethylene glycol (MEG). The rate of hydrate growth and severity of blockage formation decreased monotonically with the initial MEG fraction; visual observation of the sapphire cell suggested that MEG reduced hydrate film growth and particle deposition on the wall. The hydrate growth rate was accurately predicted within 2 vol% by a kinetic growth model. 20 gas-dominant flowloop experiments were performed with 10 vol% water flowing under annular conditions. Increasing the amount of MEG in water monotonically decreased (i) the hydrate growth rate, (ii) the rate of pressure drop increase, and (iii) the frequency of sloughing events. At subcoolings below 5 K, the pressure drop increase due to wall deposition and the global hydrate growth rate were compared. The results suggest that approximately half the hydrate was deposited on the pipeline wall; further studies are underway to expand this understanding to systems at higher subcooling. Together, these experimental studies demonstrate that a reduction in MEG fraction of up to 10 wt% from the complete inhibition requirement did not result in an imminent increase to hydrate blockage risk. The hydrate growth rate may be readily estimated through a kinetic model, where MEG fractions above 10 wt% were consistently observed to improve hydrate transportability.

400350 A New Quantitative Approach to Assess the Risk of Hydrate Blockage Formation in Subsea Flowlines

Abstract:

As production systems move toward deeper water, high operating pressures and longer thermal exposure times will significantly increase the risk of gas hydrate formation and potential flowline blockage. Over the past two decades, thermal and chemical hydrate management strategies have been developed with a goal of decreasing capital and operating costs while minimising the risk of hydrate blockage formation. The successful adoption and refinement of these management technologies hinges on accurate predictive capabilities for hydrate formation and transportability. Current simulations of hydrate blockage formation rely on a set of deterministic calculations, which do not account for distributed phenomena such as hydrate nucleation. In this presentation, we introduce a new risk engine that is designed to couple probabilistic phenomena – such as hydrate nucleation, deposit sloughing, and jamming – with existing deterministic models. The current incarnation of this engine utilises the Hydrate Flow Assurance Simulation Tool (HyFAST) to describe deterministic phenomena, where the severity of hydrate blockage formation is quantified through the increase to hydrate slurry viscosity for water- and oil-continuous systems. The first deployment of this risk engine incorporates a new probability distribution to describe hydrate nucleation as a function of system temperature; thousands of hydrate nucleation events were observed in an Automated Lag Time Apparatus (ALTA) to inform this distribution. The results of this pairing demonstrate a new ability to quantify the risk of observing hydrate blockage formation in a subsea flowline. In this presentation, we will apply this risk engine to a hypothetical long-distance tieback. The results demonstrate how flow assurance engineers can decide upon an acceptable risk tolerance (e.g. 90% of observations) and deploy this simple calculation strategy to optimise subsea design decisions, such as insulation thickness, flowline diameter, and choke opening.

400359 Prediction of Asphaltenes Solubility in Organic Solvents Via NRTL-SAC and COSMO-SAC

Abstract:

Asphaltenes are polyaromatics with hetero atoms (N, O, and S) that contribute to the polar characteristics of hydrocarbon mixture, i.e., crude oil. Solubility of asphaltenes reflecting the characteristic nature of mixture of compounds collectively, i.e., hydrophobicity, polarity, and hydrophilicity, greatly affects the production and transportation of hydrocarbon mixture, and causes fouling in unit operations. Current solubility models are based on the Scatchard–Hildebrand concept, and further extended with the Flory–Huggins correlation. These model require molecular weight and molar volume as input parameters. However, asphaltenes are solubility class rather than pure components, and the uncertainty of distribution of pure components within asphaltenes hampers the efficacy of these models. In the present work, 'like dissolves like' phenomena and solvation thermodynamics are used to calculate the activity coefficient of asphaltenes and solubility subsequently. Also, the conceptual segments of NRTL-SAC model for asphaltenes from published solubility data in mixed solvents is identified to generate the apparent sigma profile, i.e., a histogram of charge density distribution over the molecular surface of asphaltene. The sigma profile is then used in COSMO-SAC model to calculate activity coefficient and solubiliy. By comparison, the sigma profiles of asphaltenes is also obtained from quantum chemistry calculation, i.e., density functional theory, with molecular structure for asphaltene molecules as input collected from literature proposed by different authors. Furthermore, various entropy models are investigated since both Flory-Huggins and Staverman-Guggenheim models have been found with issues.

400387 Design for Mechanical Integrity in Thermal Mixing Points

Abstract:

Design for mechanical integrity in thermal mixing points

Authors: *Santhosh K Shankar**, *Gonghyun Jung*^, *Karl Anderson**, *Greg Bethke**

Thermal fatigue can occur near mixing points where fluids of different temperatures comingle to reach a desired temperature for downstream processing requirements or in hydro-processing units where hydrogen is injected to treat a hydrocarbon stream at a different temperature. Failures due to thermal fatigue may lead to loss of primary containment, resulting in fire, health hazards and production deferment resulting from unplanned unit shut-downs and are hence associated with problems of high business impact. Thermal sleeves, quills and/or spray nozzles may be used in mix points to protect the walls of the piping against fatigue. The primary challenge towards design of thermal sleeve is- understanding the process of thermal mixing near and downstream of the mixing point under different flow regimes. Identification of the appropriate dimensional and non-dimensional quantities on which the thermal mixing process depends is hence critical. The present work showcases an attempt to identify key universal parameters that govern the thermal mixing process.

Towards this objective, a parametric study was conducted using computational fluid dynamic (CFD) simulation at varying process conditions to understand the thermal behavior of the flow. Temperature statistics are probed along the wall at downstream and azimuthal locations of the pipe to map trends in the thermal behavior of the flow.

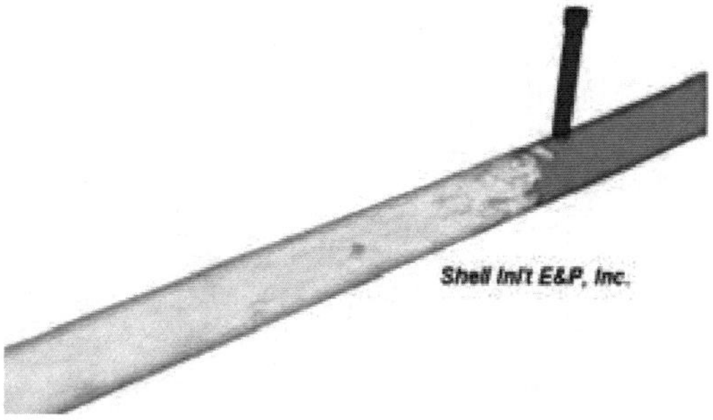

Figure: Temperature fluctuations along the pipe walls encountered in a mixing point from mixing of hot (red) and cold (blue) fluids.

399581 Stimuli-Responsive Rheoreversible Fracturing Fluids for Unconventional Oil and Gas Extraction

Abstract:

Even the most pessimistic US Energy Information Administration (EIA) models predict a near-doubling of U.S. production of tight oil by 2035 as crude production from other sources remains static or diminishes. As a result, the development of specialized fracturing fluid systems has been a constant research and development focus driven by the strong need for optimization of fracture propagation and permeability enhancement. To optimize and lessen environmental impact of reservoir stimulation, an environmentally benign, CO2-activated, rheoreversible fracturing fluid is presented. This fracturing fluid enhances permeability due to in situ volume expansion triggered by the presence of CO2. Our results show **100 % volume expansion** of PNNL fracturing fluids at temperatures relevant to tight oil reservoirs. This rapid volume expansion created fracture networks in highly impermeable crystalline rock cores with **permeability values up to 3 orders of magnitude higher** than the permeabilities obtained when using conventional fracturing fluids. The chemical mechanism, phase-change behavior, and rheology of this fracturing fluid at tight oil temperatures and pressures are presented.

Mitigating Downhole Flow Assurance Issues Using Specialist Modeling Software

Fabin Xu[1], Chunmin Zeng[1], Chong Zhang[1], Zhao Dong[1], Xiaohong Zhang[2], Adrian Ferramosca[3], Behnam Salimi[2] and Richard Szczepanski[2]

(1) CNOOC Zhanjiang Exploration and Development Research Institute, Zhanjiang, China
(2) Infochem PVT Physical Property and Thermodynamics, KBC Advanced Technologies plc, London, United Kingdom
(3) Weatherford, Edinburgh, United Kingdom

We are currently exploiting a large oilfield in the Beibu Gulf, South China Sea, requiring the design of 12 wells. We are using specialized well-modeling software using advanced flow assurance techniques incorporating accurate wax, asphaltene and hydrate prediction methods to revise our previous design to include additional solids deposition mitigation techniques, including thermal insulation in a number of wells where this was not originally thought to be a problem.

Fluid Description

Source Composition

Fluid Component	MolarComposition (Fraction)
H2O	0.0098
N2	0.0108
H2S	0.0000
CO2	0.0302
C1	0.5103
C2	0.0244
C3	0.0616
IC4	0.0059
NC4	0.0124
IC5	0.0483
NC5	0.0000
C6_1	0.2442
C6_2	0.0419

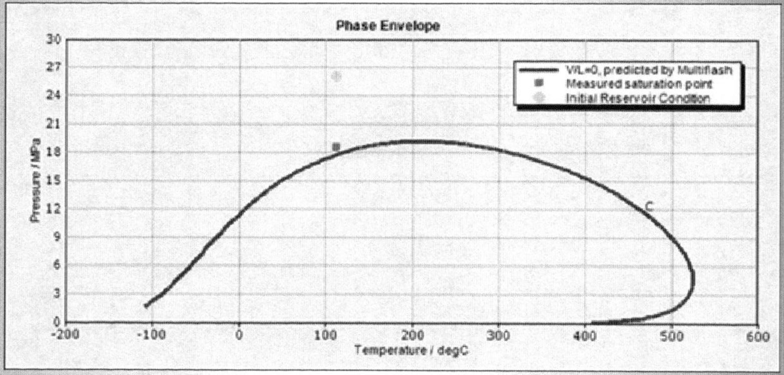

Software used

WellFlo™ v6.0 (Weatherford)

PVTFlex™ v2.0 (Weatherford)

Multiflash™ v4.4 (KBC-Infochem)

Weatherford's WellFlo™ well design and evaluation software, through its direct link to KBC-Infochem's Multiflash™ rigorous wax, hydrate and asphaltene flow assurance models, is able to simultaneously plot the solids formation phase envelope together with the pressure-temperature profile followed by the hydrocarbon fluids from bottom-hole to wellhead to platform.

WellFlo™ Plots

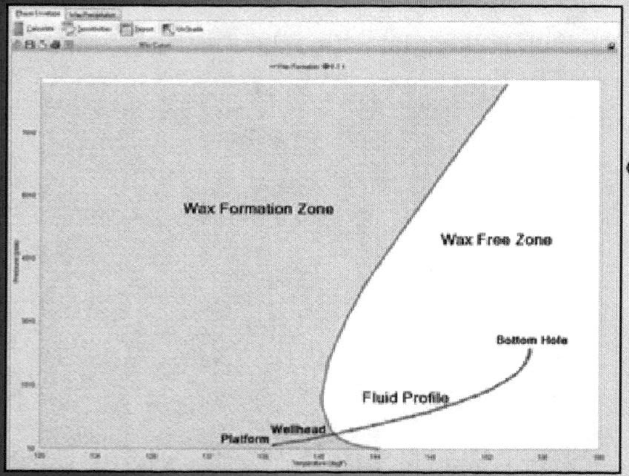

The plot on the left shows the predicted pressure – temperature profile of the fluids from bottom-hole to surface, showing how the wax-producing zone is entered toward the top of the wellbore.

Adding insulation to the design enables this profile to avoid the wax formation zone as shown in the plot on the right.

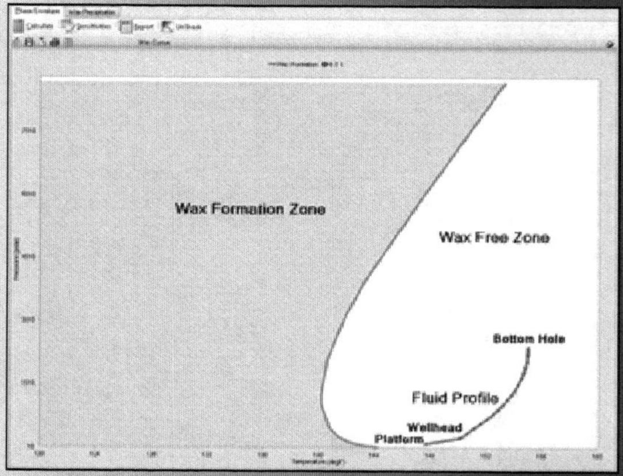

Interaction of Oil and Gas Shale with Drilling Muds and Fracturing Fluids

Samyukta Koteeswaran, Peter E. Clark

Department of Chemical Engineering, Oklahoma State University

Introduction

The boom in drilling and completions in shale formations has been partially responsible for the large increase in oil production in the United States [1]. To a large extent, development of shale resources has been facilitated by the use of high-rate fracturing treatments. These treatments typically use friction reducers to lower surface pressure requirements, and thus job costs. Probably the most common polymers used are polyacrylamides. While this family of polymers provides excellent friction reduction, they often interact with shales in a manner that can potentially cause formation damage problems. The polymer – shale interactions are well known and have been used for years to stabilize shales during drilling. Since shales typically have low permeability, it is difficult to ascertain the magnitude of formation damage by polyacrylamide based fracturing fluids.

The alteration of rock properties such as wettability and permeability is a common drilling problem [2]. Oil production will be restricted if the permeability is lowered. A change in wettability will have serious implications for both the core analysis and for formation damage, wettability also has a major impact in multi-phase flow in porous media [2]. The components that are blended with drilling mud are acceptable if they do not alter the rock properties. However, all the additives alter wettability and significantly reduce the permeability. Additives such as surfactants, clays, polymers and other chemicals could interact with the formation and could cause formation damage. Formation damage is caused by invasion of foreign fluids near the well bore [3, 4]. The study of rock mechanics and chemical interactions between shale and drilling mud is important before pumping drilling mud into the well as drilling fluids could considerably reduce the permeability [5].

Inhibited fluid systems that use polyacrylamides have proven effective for drilling through shale. Good shale inhibitors limit ion exchange into interlayer areas and prevent hydration of surfaces. The additives are adsorbed to the clay surface by hydrogen bonding to raise the zeta potential and thereby strengthen the clay gel structure [6, 7]. Shale stability can be achieved by combining potassium chloride with partially hydrolyzed polyacrylamides [8]. Potassium chloride is added to the mud system to reduce the swelling tendency of water sensitive shales.

Shales in each region behave differently when exposed to drilling fluid, polyacrylamides in particular. The interaction of the polyacrylamides with shales depends on number of factors such as swellability, permeability, surface charge, porosity and mineralogy of shales. The shale should be analyzed before injecting it with the drilling fluid. The mineral composition of the shales in this study is determined using SEM - EDS techniques. Surface characteristics and surface charge of the shale is analyzed using Scanning Electron Microscopy and Zeta Potential Analysis respectively. The interaction of polyacrylamides with the shales is quantified by measuring the rheology of shale slurries with polyacrylamides.

In this study the interactivity of shale with polyacrylamides is quantified and it is correlated to the surface properties of the shale.

Materials and Methods

Sample Preparation

Polymer

Anionic and Cationic Polyacrylamide (PAM) of average molecular weight of 10^7 g/gmol is used. Kemira supplied the polyacrylamide samples. The sample is measure by weight and added to DI water and left in shaker table at a speed of 200 RPM for 15 minutes. The sample is then left to hydrate at room temperature for 24 h.

Shale Sample

New Albany, Chattanooga, Gorgas and Woodford shale samples are cut using a coresaw from the core samples. The shale is ground using a belart mixer and sieved to obtain particles less than 75 micrometer. These samples are measured by weight and added to DI water and left in shaker table for 30 minutes at speed of 500 RPM. The sample is left to hydrate for 24 h at room temperature before adding it to the polymer sample.

Equipment

Rheometer

The Discovery HR-3 (DHR-3), advanced single head controlled stress rheometer is used for making rheological measurements of the sample. Rheometer is a highly sensitive device for probing molecular structure. Parallel plate geometry with gap from 0.5 mm to 1 mm is used for samples with viscosities >10 mPa.s. Double wall concentric cylinder is used for measuring the properties of low viscous samples. All the experiments were done at a temperature of 25 °C ± 0.03 °C. Three types of test were conducted on each sample. They are flow ramp test, frequency sweep, and amplitude sweep. Flow ramp tests are done for shear rates from 0.001 s^{-1} to 200 s^{-1}. The time is chosen depending on the viscosity of the samples. Frequency sweep test is done from 0.1 rad/s to 20 rad/s at 5% strain. Amplitude sweep is done from 0.1 – 100% strain at 10 rad/s. Pre shearing of all the samples for 30 s at 200 s^{-1} is done before each test. Amplitude sweep test is done to determine the strain % for the frequency sweep test.

Geometry

To make rheological measurements of the slurries vane geometry is used. The use of vane technique prevents wall slippage and also particle settling during the measurements.

Zeta Potential Analyzer

Zeta PALS model made by Brookhaven Instruments Corporation is used for measuring the net surface charge of the shale particles. The experiment was conducted and room temperature and repeated thrice to get concurrence.

Analysis

Model

Carreau Model was used to determine the zero shear rate viscosity for the entire polymer – shale and polymer samples. This five-parameter model has sufficient flexibility to fit a wide variety of experimental curves.

This viscosity model allows data to be fitted to the following model

$$\frac{\eta - \eta_\infty}{\eta_0 - \eta_\infty} = \frac{1}{(1 + (K_{1\gamma})^2)^{n^{1/2}}}$$

Where η_0 the Newtonian viscosity, η_∞ the infinite viscosity, K the consistency and n the power law index.

This model is used to fit data from a fluid that have a lower Newtonian and upper Newtonian region with a power law region in between. The Carreau model is a variant to Ellis model and an alternative to the Cross model. The Carreau model attempts to describe a wide range of fluids by the establishment of a curve-fit to piece together functions for both non-newtonian and shear thinning ($\eta < 1$) non-Newtonian laws.

Rheology Tests

Flow curve and Step Shear Rate

The polyacrylamide – shale sample is subject to continuous step shear rate from 0.001 s^{-1} to 200 s^{-1} . The flow ramp test is repeated thrice to get good concurrence. Flow ramp test is used to study the behavior of the material over a spectrum of shear rates. The zero shear viscosity of all the samples is determined by fitting the Carreau model using the inbuilt TRIOS software.

Scanning Electron Microscopy

SEM techniques were used to study the surface properties and morphology of the shale under study. Shale core were core sawed to get 1 – 2 mm shale particles, parallel to the bedding plane. The sample was placed on the stub and the shales being non conducting materials were sputter coated with conducting layers of carbon. The surface of the shale was examined using different magnifications. In order to determine the elemental composition of shales, the shale samples were coated with layers of carbon and Energy Dispersive Spectroscopy (EDS) analysis of the shale samples were done using Evex nanoanalysis software in SEM.

Results and Discussion

In this section we would be discussing the results of the rheology of shale samples with anionic and cationic polyacrylamide. The shale mineralogy, surface properties of shale and interaction of shale with the polyacrylamides is measured quantitatively. Table 1. Shows the mineralogy of shales under study.

Table 1: Mineralogy of shale samples determined using SEM – EDS

Elements	New Albany (Wt. %)	Gorgas (Wt. %)	Chattanooga (Wt. %)	Woodford (Wt. %)
Al	13.43	15.49	11.21	15.59
Si	27.39	58.98	76.71	67.38
Ti	0.74	1.4	1.13	-
S	19.77	-	0.36	-
K	9	12.76	7.95	10.54
Fe	29.68	11.36	2.63	4.3
Mg	-	-	-	2.2

Table 2. Shows the surface charge for each shale determined using the Zeta Potential analysis. It is clear that the net surface charge in all the shales under study is highly anionic in nature. The surface charge of the shale does not have a great influence on the interaction of polyacrylamide with shales but higher the net negative charge of shale lower is its interaction with anionic polyacrylamide.

Table 1: Zeta Potential of Shales

Shale	Zeta Potential (mV)	pH
New Albany	-28.45	8.12
Gorgas	-34.39	8.73
Chattanooga	-23.69	7.48
Woodford	-28.58	7.78

The interaction of polyacrylamide with shale is measured rheologically. In this method, the concentration of the shale is kept constant at 0.5 lb/bbl and the anionic polyacrylamide concentration is varied from 0.1 wt. % to 0.3 wt. %. This is the concentration range within which the polyacryalmides are used in drilling muds. Figure 1. Shows the zero shear rate viscosity change of anionic polyacrylamide – shale slurry for increasing concentration of anionic polyacrylamide. From the figure it could be seen that Chattanooga shale has a bigger increase in viscosity with increasing anionic polyacrylamide concentration and Gorgas shale has the least increase in viscosity. This clearly shows that each shale interacts differently with polyacrylamides.

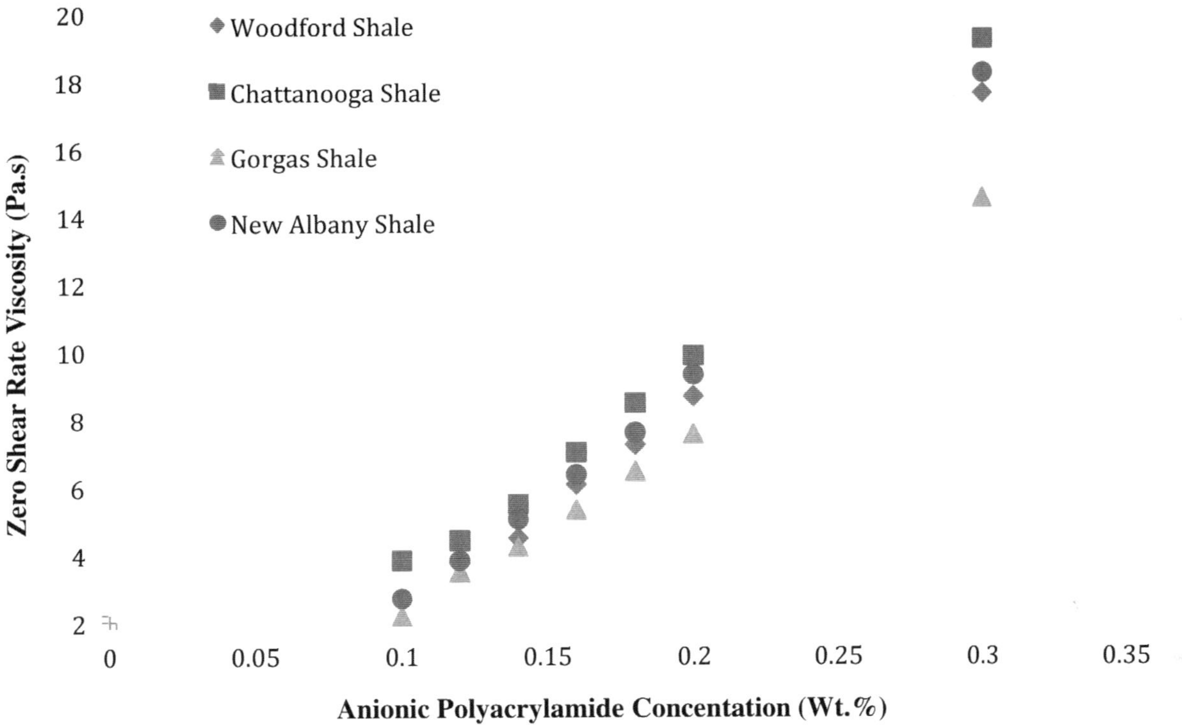

Figure 1: Zero Shear Rate Viscosity (Pa.s) vs Anionic Polyacrylamide Concentration (Wt. %) of the shale samples

Conclusion

Polyacrylamides are most commonly used friction reducers and the interaction of polyacrylamide shale interaction is not widely studied. In this paper, we have proposed a method to quantitatively measure the polyacrylamide – shale interaction. Each shale interacts differently with anionic polyacrylamide, with Chattanooga shale having highest interaction and Gorgas shale the lowest interaction with anionic polyacrylamide among the four shales under study. It is imperative that the shale – fluid interaction is studied for each shale formation before injecting the fluid.

Acknowledgements

We would like to thank the following organizations and people who helped us during the course of work

- Our funding source, Research Partnership to Secure Energy for America.
- Dr. Jack C. Pashin, School of Geology, Oklahoma State University.
- Dr. Jim Puckette, School of Geology, Oklahoma State University.
- Dr. Josh Ramsay, School of Chemical Engineering, Oklahoma State University.
- Department of Chemical Engineering, Oklahoma State University.

References

1. Shah, S.N., N. Shanker, and C. Ogugbue. *Future challenges of drilling fluids and their rheological measurements*. in *AADE fluids conference and exhibition, Houston, Texas*. 2010.
2. Sharma, M.M. and R.W. Wunderlich, *The Alteration of Rock Properties Due to Interactions With Drilling Fluid Components*. Society of Petroleum Engineers.
3. Mahto, V. and V.P. Sharma, *Rheological study of a water based oil well drilling fluid.* Journal of Petroleum Science and Engineering, 2004. **45**(1–2): p. 123-128.
4. Urbissinova, T., J.J. Trivedi, and E. Kuru, *Effect of Elasticity during Viscoelastic Polymer Flooding - A Possible Mechanism of Increasing the Sweep Efficiency*. Society of Petroleum Engineers.
5. Corr, C.C. a, and R.S.V. Nascimento, *Study of shale-fluid interactions using thermogravimetry.* Journal of Thermal Analysis and Calorimetry, 2005. **79**(2): p. 295-298.
6. Guo, W.-Y. and B. Peng, *Synthesis and characterization of starch-graft-polyacrylamide copolymers and their application as filtration control agents in drilling fluid.* Journal of Vinyl and Additive Technology, 2012. **18**(4): p. 261-266.
7. Han, M., et al., *Laboratory Investigation of the Injectivity of Sulfonated Polyacrylamide Solutions into Carbonate Reservoir Rocks*. Society of Petroleum Engineers.
8. Clark, R.K., et al., *Polyacrylamide/Potassium-Chloride Mud for Drilling Water-Sensitive Shales.*

400443 Evaluation of the Methods for Predicting Dynamic Viscosity of High-Temperatures and High-Pressures Deep Petroleum Formations

Abstract:

Ten methods of predicting dynamic viscosity for high-temperatures and high-pressures (HTHP) deep-water petroleum formations are evaluated for ease of implementation and applications for flow assurance modeling, pipeline flow simulation and reservoir simulation of carbon dioxide sequestration in the depleted oil-reservoirs. The evaluated techniques include various versions of the *dead-oil viscosity correlation*, the *Stiel-Thodos viscosity correlation*, the *Little-Kennedy correlation*, the *free-volume viscosity theory*, the *frictional theory for viscosity*, the *kinetic theory description for viscosity*, the *Van der Waals transport equations-of-state*, the *Ely-Hanley corresponding states viscosity methods*, the *hard-sphere theory of viscosity by Dymond-Assael*, the *Yarranton-Satyro viscosity correlations* and those versions of viscosity methods that are coupled with the modified Tait equation, VDW EoS, CPA EoS and SAFT+ cubic EoS for liquid density estimations at the HTHP conditions. The HTHP dynamic viscosity data used in the evaluation of the methods for predicting dynamic viscosities are reported in the literature by Enick Group at U of Pitt, Caudwell-Trusler-Wakeham Group at Imperial College London and Ducoulombier et al. in France. While all the evaluated methods gave various degrees of uncertainties at HTHP conditions (due to the judicious choice of the inaccurate liquid density correlations), the VDW viscosity equation of state (which does not require liquid densities or volumetric properties as input data) performs better and in agreement with the measured dynamic viscosities at the HTHP conditions of deep-water petroleum formations. It also provides a very suitable mean of predicting viscosity over the entire PVT states of HTHP densities, including dilute, dense-fluids, supercritical, near-critical and at critical viscosity. This article is useful for all type of analyses of flow assurance modeling.

Keywords: cubic equation, viscosity methods, dynamic viscosity, volumetric property, flow assurance

400463 The Van Der Waals Ten Commandments for Cubic Equations-of-State (Electronic Poster Presentation)

Abstract:

The reading of the Van der Waals (VDW) 1910 Nobel Prize Lecture[1,2] reveals the hindsight and foresights of the VDW theory of cubic equations of state for the individual pure substances and mixtures. A correct reading of the VDW Nobel Prize Lecture reveals some Do's and Don'ts and their consequences led the author to the following itemized Van der Waals Ten Commandments for Cubic Equations-of-State, which begins with Oh the Faithful:

- Be Loyal to the VDW Cubic Form of Attractive and Repulsive Expressions
- Be Loyal to the VDW Ultimate Objective: Construct A Substance-Based Cubic Equation
- Be Loyal to the VDW Asymptotic Critical Volume-Limit: Weak-Point of VDW Theory
- Be Loyal to the Four Properties of VDW Theory of Cubic Equations of State
- Be Loyal to the VDW Empirically-Based Molecular Parameters for Reformed Equations
- Be Loyal to the VDW Critical Point as Limit of Phenomenological Gas-Liquid Transition
- Be Loyal to the VDW Defined Gas-Constant for Unifying Caloric and Transport EOS

- Be Loyal to the VDW Continuity of Gas and Liquid States for Temperature Functions
- Be Loyal to the VDW Constraining Temperature Parameters to a(T) and b(T)
- Be Loyal to the VDW Corresponding States Principles for Generalized Property Charts

1. INTRODUCTION AND OBJECTIVE

We recently celebrated the hundred year anniversary[1,2] of the Nobel Prize Lecture delivered by J. D. Van der Waals on December 10, 1910 entitled "*The Equation of State for Gases and Liquids*."

This poster highlights the salient points, including the obvious and non-obvious facts about the author's reading of the 1910 Nobel Prize Lecture. In particular, areas of the Nobel Lecture where noticeable quotes have emerged over the years into textbooks, seminars, conferences and symposium are addressed from the viewpoint of the author's experience. The objective of the poster is to highlight fact that can be applied in the future development of the Van der Waals cubic equations of state.

2. DISCUSSION OF THE TEN COMMANDMENTS

It is revealed in this poster that the difficulties encountered with the modified VDW type of cubic equations can be traced to the violation of one or more of those VDW Commandments: as for instance the lack of *continuously* differentiable temperature functions a(T) and b(T) leads to difficulties in the supercritical regions or to the infinite isochoric heat capacity at the critical point or to the impossibility of simultaneously predicting accurate vapor pressure and virial coefficients with the same functional form of temperature-dependent parameters.

Applying the VDW gas constant, R^{VDW} provides a way to unite thermodynamic and transport equations of state and also it is a way for the direct use of the VDW cubic equations of state for predicting thermodynamic properties as opposed to the widely accepted route of the PVT derivatives: after all the PVT derivatives are based upon the temperature functions and ***not*** the VDW parameters a (T_c) and b (T_c).

Also, it is by using empirically-based physical parameters in the reformed VDW 1873 equation that would stop the proliferation of the multiplicity of 2-P and 3-P cubic equations of state. VDW theory is a modified form of the ideal gas-law with molecular-based empirical parameters for correcting the incomplete VDW theory.

Furthermore, those investigators that are passionately advocating the replacement of VDW repulsive term forget that the ***theory of liquids*** of the VDW theory is based solely on the VDW repulsion term as shown by the following limiting behavior:

$$\lim_{\substack{P\to\infty \\ \text{for all } T}} v \to b(T_c)$$

But, in accordance with the VDW theory, the Z_c of cubic equation ***solely*** depends on the ratio of b/v_c, as shown by

$$Z_c = \frac{1-2\left(\frac{b}{v_c}\right)}{2\left[1-\left(\frac{b}{v_c}\right)\right]^2}$$

Perhaps knowing that fact, VDW stated in the 1910 Nobel Prize Lecture[2]: "the weak point of my theory is b," so whenever $b/v_c = 1/3$ (or $v_c = 3b$), $Z_c = 3/8$ of the VDW 1873 cubic equation of state.

By the way, before we agree with the notion that the VDW theory shows ***anomaly*** behavior at the fluid critical point, we need to examine the adherence to the following limiting critical behavior that should be imposed on all the VDW cubic equations of state:

$$\lim_{\substack{T\to T_c\\ P\to P_c}} v \to v_c \quad \text{and} \quad \lim_{\substack{T\to T_c\\ P\to P_c}} Z \to Z_c$$

That stipulated physical boundary condition states that the VDW cubic equations of state should predict at fluid critical point: accurate ***critical volume*** and ***critical compressibility factor*** of the individual pure substances or mixture of fixed overall composition. In that case, the parameters of cubic equations of state should be constrained by ***four*** critical constraint criteria because there are *four properties* of the VDW 1873 equation: Z_c =3/8; Ω_a = 27/64; Ω_b = 1/8; Ω_w (or b/v_c) = 1/3 and that is in accord with the theory of cubic polynomial equations. Thus, *four unrelated* parameters are required in the VDW 1873 cubic equation; simply, parameters a and b are necessary but ***insufficient*** to resolve the VDW 1873 cubic equation at fluid critical point); the required *four* critical constraint criteria expressions are stated as:

$$\left(\frac{\partial P}{\partial v}\right)_T = 0; \quad \left(\frac{\partial^2 P}{\partial v^2}\right)_T = 0$$

$$f(P_c, v_c, T_c) = 0; \quad (v - v_c)^3 \equiv (Z - Z_c)^3 = 0$$

Besides the combining rules for parameters a_m and b_m that VDW introduced in 1888, the justification for more composition-dependent parameters in the VDW 1873 cubic equation is based on the *insufficiency* of the expressions for the critical properties:

$$P_c = \frac{1}{27}\frac{a_m}{b_m^2}; \quad T_c = \frac{8}{27}\frac{a_m}{Rb_m}; \quad v_c = 3b_m$$

While P_c, v_c and T_c vary with composition of mixture in the expression for the critical properties, the Z_c does not vary with composition, being identically the same as for the pure substance; that is the major source of errors in predicting critical properties from the VDW 1873 equation. Consequently, additional parameters are required to be embedded in the reformed VDW 1873 equation to express $Z_c = f$ (mixture composition). That is another justification for the design of the four-parameter Lawal-Lake-Silberberg equation of state.

The statistical mechanical derivation of the virial coefficients from the VDW 1873 equation is reported by Hill (1947, 1948)[3] as inaccurate after the second virial coefficient; as seen from the virial expansion of the VDW 1873 cubic equation,

$$Z = 1 + \left(b(T_c) - \frac{a(T_c)}{RT}\right)\frac{1}{v} + \left(\frac{b(T_c)}{v}\right)^2 + \left(\frac{b(T_c)}{v}\right)^3 + \left(\frac{b(T_c)}{v}\right)^4 + \cdots$$

The virial expression shows that the attractive parameter a (T_c) ***disappears*** from the *third and higher* virial coefficients: that is ***not*** VDW theory because the VDW concept is $Z_{rep} + Z_{att}$. Therefore, the remedy is for more parameters in denominator of a/v^2 term (as the theory of cubic polynomial equations stipulates *four unrelated* parameters for cubic equations), which was understood by Clausius (1880) and Berthelot (1900) in their reformed VDW cubic equations of state. However, it is not always obvious from the coefficients of the virial expansion of the VDW 1873 equation that the second virial is the *building blocks* for higher virial coefficients; the fact is shown by the following limiting values of PVT data for the second (B), third (C) and fourth (D) virial coefficients:

$$B = \lim_{1/v\to 0} (Z - 1)v$$

$$C = \lim_{1/v\to 0} [(Z - 1)v - B]v$$

$$D = \lim_{1/v\to 0} [(Z - 1)v^2 - Bv - C]v$$

Those limiting values of PVT data uniquely justified more parameters in the denominator of the a/v^2 term of the VDW 1873 equation of state because the expression of the second virial coefficient (B) requires parameters $a(T_c)$ and $b(T_c)$ as the building blocks for the third and higher virial coefficients: that is another justification for the construction of the four-parameter Lawal-Lake-Silberberg equation of state.

By predicting physical properties using the VDW theory of cubic equations of state, the author agrees with the statement by Guggenheim (**1945**)[4] that the corresponding states principle "may safely be regarded as the most useful by-product of the Van der Waals 1873 equation of state." Therefore, besides the temperature-dependent parameters a (T) and b (T), no other parameters embedded in the reformed VDW cubic equations of state should depend on temperature, otherwise the *meaning* of the Law of Corresponding States (LCS) is ambiguous. Hence, temperature-dependent binary interaction parameter is meaningless in the context of the LCS. The design of cubic equations of state reported by Himpan (1951) and Heyen (1980, 1981), having more temperature-dependent parameters than stipulated by the LCS, violates the corresponding states principle.

3. CONCLUSION

By adherence to the unspoken words of Van der Waals, which are casted as Ten Commandments in this poster, we can resolve the major disagreement between the Van der Waals theory and fluid properties, including the fluid critical point. Evidently, there should be a particular reason to construct another VDW cubic equation of state, otherwise we would reached the point of diminishing returns and joint others in the high degree of trivialization of the VDW theory of cubic equations of state.

REFERENCES:

1. **Journal of Supercritical Fluids,** Volume 55, Issue 2, Pages 401-860 (December 2010)

100th year Anniversary of van der Waals' Nobel Lecture (Edited by Sona Raeissi, Maaike C. Kroon and Cor J. Peters)

2. Van der Waals, J. D., **1910** Nobel Prize Lecture available from "J. D. van der Waals - Nobel Lecture: The Equation of State for Gases and Liquids". http://www.nobelprize.org/nobel_prizes/physics/laureates/1910/waals-lecture.html

3. Hill, T. L., "Free-Volume Models for Liquids," J. *Phys. and Colloid Chem.*, *51*, 1219 **1947**; "Derivation of the Complete Van der Waals' Equation from Statistical Mechanics," *J. Chem. Educ.*, *25* (6), 347, **1948**

4. Guggenheim, E. A., The Principle of Corresponding States. *J. Chem. Phys*. 13, 253, **1945**

400466 Characterization of Critical Properties and Acentric Factors for High-Molecular-Weight-Fluids, Petroleum Fractions, Coal and Natural Products

Abstract:

Characterizing methods reported by Walas (1985), Poling et al. (2001) and Ihmels-Ambrose (2010) for pure substances are based on structural formula and therefore not suitable for pseudo-component properties (or poorly characterized mixtures encountered with petroleum, coal and natural products, which are usually based on measured properties such as average molecular weight, boiling point and specific gravity (at 20°C)[1] rather than molecular structure). The Paraffinic-Naphthenic-Aromatic (PNA) reported by Peng- Robinson (1978) or the group contribution methods by Lydersen et al. (1955) (or the methods based on carbon number reported by Whitson (1983) and Panuganti-Chapman (2013) are not considered in this report because the intention here is not to gain molecular insights into crude-oil properties but rather to treat pseudo-component properties as *single-lumped component* for the sole purpose of modeling by using cubic equations of state. Also, the trusted Database as tabulated by the NIST Groups for the critical properties and acentric factors do not include n-tetracontane ($nC_{40}H_{82}$), even though specific volumes at several temperatures and pressures for n-tetracontane have been reported by Arthur Doolittle since 1964 (*J. Chem. Eng. Data*, *9* (2), 275, 1964).

Consequently, thirty-six methods of characterizing critical properties and acentric factors for the parameters of cubic equations of state used in the process and reservoir simulation are evaluated for ease of implementation and use for high-molecular-weight-fluids, petroleum fractions, coal tar liquids and natural products. Analyses of results shows the methods are not adequate for poorly characterized pseudo-components of high aromatic, naphthenic and cycloalkanes;

and beyond n-Octadecane hydrocarbons, some critical property parameters (critical pressure and volume, critical-Z factor and acentric factor: which is greater than 1 and therefore, is physically meaningless) are usually based or trial-and trial error techniques: thus, lacking consistency with lighter hydrocarbons in the petroleum fractions assay. While most of the estimation methods provide satisfactory predictions of the critical temperatures, they all underestimated critical pressures of the pseudo-components.

By using the API Database as reported by the University of Pennsylvania Group and gathering data from *Hydrocarbon Processing* and other relevant sources (theses and dissertations reports), we developed a Paraffinic-Naphthenic-Aromatic Characterization Factor into the correlation for critical temperature in terms of average molecular weight (Mw), boiling point (T_b) and specific gravity (S_g) and we use several cross-plotting of the ratio of boiling point to critical temperature (T_b/T_c) versus molecular weight, critical properties (P_c, V_c) and acentric factor (ω) to generate high-correlation coefficients; correlating groups, such as (T_b/T_c), (T_c/P_c), (V_c/Z_c), (M_w/P_c), (M_w/T_c) and Z_c versus molecular weight or acentric factor show a high-correlation coefficients. While we claim no theoretical basis (apart from the high-correlation coefficients and the statistically derived predictive groups), the principles of dimensional similitude provide the convenient basis in the selected groupings. The grouping parameters reduce the number of tailored-fitting parameters and thus preclude the need for extensive experimentally measured data for the unstable petroleum fractions and pseudo-components.

Hundred petroleum fractions reported by Roess (1936), Lenoir-Hipkin enthalpy (1973) and petroleum fraction data are used to validate the pseudo-component critical properties developed in this reported. In comparison with available correlative methods, the current approach is better in accuracy and ease of implementation in cubic equations of state modeling. The current approach is also validated for predicting viscosity of coal liquids (Baltatu, 1985; 1995) as well as predicting densities of natural products. While the current approach may not be convenient for the type of work in the petroleum refining operations, the methodology is superior for empirical characterization of pseudo-component critical properties and acentric factors as lumped-single component and it thus offers another technique to the traditional *pseudo-component*, or *hypothetical-component* modeling approach for process and reservoir simulation models. It is therefore recommended for high-molecular-weight components for which experimental critical properties are typically unavailable.

References:

1. Riazi, R. M. *Characterization and Properties of Petroleum Fractions*; ASTM: Philadelphia, **2005**

2. API, *Properties of Hydrocarbons of High Molecular Weight*, API Research Project 42, American Petroleum Institute, Washington, DC, **1966**.

3. Riazi, M. R., Daubert, T. E., "Simplify Property Prediction," *Hydrocarbon Process, Petrol. Refiner*, *59*(3), 115-116, **1980**

4. Lin, H. M., Chao, K. C., "Correlation of Critical Properties and Acentric Factor of Hydrocarbon Derivatives" *AIChE J.*, *30*(6), 153-158, **1984**

5. Watson, K. M., Nelson, E. F., Murphy, G. B., "Characterization of Petroleum Fractions," *Ind. Eng. Chem. 27*, 1460-4, **1935**.

6. Winn, F. W., "Simplified Monograph Presentation, Characterization of Petroleum Fraction," *Pet. Refin.*, *36*(2), 157, **1957**

7. Smith, R. L., Watson, K. M., "Boiling Points and Critical Properties of Hydrocarbon Mixtures," *Ind. Eng. Chem.*, 29, 1408, **1937**

8. Watansiri, S., Owens, V. H., Starling, K. E., "Correlation for Estimating Critical Constants, Acentric Factor, and Dipole Moment for Undefined Coal-Fluid Fractions," *Ind. Eng. Chem. Process Des. Dev.*, *24*, 294-96, **1985**

9. Riazi, M. R., Daubert, T. E., "Characterization Parameters for Petroleum Fractions," *Ind. Eng. Chem. Res*., *26*(24), 755-759, **1987**

10. Abbott, M. M., Kaufmann, T. G., Domash, L., "A Correlation for Petroleum Liquid Viscosities of Petroleum Fractions," *Can. J. Chem. Eng., 49,* 379, June, **1971**

11. Riazi, M. R., Al-Otaibi, G. N., "Estimation of Viscosity of Liquid Hydrocarbon Systems," *Fuel 80*, 27-32, **2001**

12. Cullick, A. S., Pebdani, F. N., Griewank, A. K., "Modified corresponding states method for predicting densities of petroleum reservoir fluids," *Ind. Eng. Chem. Res*., *28* (3), 340-47, **1989**

13. Moharam, H. M., Fahim, M. A., "Prediction of Viscosity of Heavy Petroleum Fractions and Crude Oils Using a Corresponding States Method," Ind. *Eng. Chem. Res*., *34* (11), 4140-44, **1995**

14. Sharma, R., Sunil Goel, S., "Correlation of Viscosity of Coal Liquids," *Ind. Eng. Chem. Res*., 36 (9), 3999-4007, **1997**

15. Baltatu, M. E., "Prediction of the liquid viscosity for petroleum fractions," *Ind. Eng. Chem. Proc. Des. Dev*., *21* (1), 192-5, **1982**

16. Baltatu, M. E., Ely, J. F., Hanley, H. J. M., "Thermal Conductivity of Coal-Derived Liquids and Petroleum Fractions," *Ind. Eng. Chem. Process Des. Dev*., *24* (2), 325, **1985**

17. Baltatu, M. E., Chong, R. A., Huber, M. L., Laesecke, A., "Transport Properties of Petroleum Fractions," *Int. J. Thermophysics*, *20* (1), 85-95, **1999**

18. Jacoby, R. H., Yarborough, L., "PVT Measurements on Petroleum Reservoir Fluids and Their Uses," *Ind. Eng. Chem*., *59*(10), 49, **1967**

19. Twu, C. H., "An Internally Consistent Correlations for Predicting the Critical Properties and Molecular Weights of Petroleum and Coal-Tar Liquids," *Fluid Phase Equil*., *16*, 137, **1984**

20. Allen, J. C., "A Review of the Properties of Some Oil and Gas Reservoirs," *Oil and Gas J*., *51*(12), 355, **1952**

21. Cragoe, C. S., "Thermodynamics of Petroleum Products," *Bureau of Standard, U. S. Department of Commerce, Miscellaneous Publication*, No. 97, 22, **1929**

22. Roess, L. C., "Determination of Critical Temperature and Critical Pressure of Petroleum Fractions by a Flow Method," *J. Soc. Pet. Tech*. (London), *22*, 665, **1937**

23. Lenoir, J. M., Hipkin, H. G., "Measured enthalpies of eight hydrocarbon fractions," *J. Chem. Eng. Data*, *18* (2), 195-202, **1973**

24. Moharam, H. M., Braek, A. M., Fahim, M. A., "A Simple Method of Estimating the Enthalpy of Petroleum Fractions," *Ind. Eng. Chem. Res*., *37* (12), 4898-4902, **1998**

25. Filipe A. Z., Da Silva Lopes, V., Kontogeorgis, G. M., "Approach Suitable for Screening Estimation Methods for Critical Properties of Heavy Compounds," *Ind. Eng. Chem. Res*., *45* (1), 476-480, **2006**

26. Jianzhong, Z., Biao, Z., Suoqi, Z., Renan, W., Guanghua, Y., "Simplified Prediction of Critical Properties of Nonpolar Compounds, Petroleum, and Coal Liquid Fractions," *Ind. Eng. Chem. Res*., *37* (5), 2059-60, **1998**

27. Brule, M. R., Lin, C. T., Lee, L. L., Starling, K. E., "Multiparameter corresponding-states correlation of coal-fluid thermodynamic properties," *AIChE J*., *28* (4), 616-25, **1982**

28. Wilson, G. M., Johnston, R. H., Hwang, S. C., Tsonopoulos, C., "Volatility of coal liquids at high temperatures

and pressures," *Ind. Eng. Chem. Proc. Des. Dev.*, *20* (1), 94-104, **1981**

29. Tsonopoulos, C., Heidman, J. L., Hwang, S. C., *Thermodynamic and Transport Properties of Coal Liquids*, New York, **1986**

30. Whitson, C. H., "Characterizing hydrocarbon plus fractions," *Soc. Pet. Eng. J. 23*, 4, 683-694, **1983**

31. Panuganti, S. R., Vargas, F. M., Chapman, W. G., Property Scaling Relations for Nonpolar Hydrocarbons, *Ind. Eng. Chem. Res.*, **2013**, *52*, 8009-802

32. Lydersen, A. L., Greenkorn, R. A., Hougen, O. A., *Generalized Thermodynamic Properties of Pure Fluids*, University of Wisconsin, Engineering Experimental Station, Report 4, Madison, Wisconsin, October, **1955**

33. Lydersen, A. L., "Estimation of Critical Properties of Organic Compounds by the Method of Group Contributions," University of Wisconsin, Engineering Experimental Station, Report 3, Madison, Wisconsin, April, **1955**

34. Christian Ihmels, E., "The Critical Surface," *J. Chem. Eng. Data*, *55* (9), 3474-3480, **2010**

35. Sovova, H., Galushko, A. A., Stateva, R. P., Rochova, K., Sajfrtova, M., Bartlova, M., "Supercritical fluid extraction of minor components of vegetable oils: β-Sitosterol," *Journal of Food Engineering*, *101*(2), 201-209, **2010**

36. Freitas, S.V.D., E Silva, F.A., Pastoriza-Gallego, M. J., Piñeiro, M. M., Lima, A. S., Coutinho J. A. P., "Measurement and prediction of densities of vegetable oils at pressures up to 45 MPa," J. Chem. Eng. Data, 58 (11), 3046-53, **2013**

37. Poling, B. E., Prausnitz, J. M., O'Connell, J. P., *The Properties of Gases and Liquids*, (5th ed.) McGraw-Hill, New York, **2001**

38. Walas, S. M., *Phase Equilibria in Chemical Engineering*, Butterworth Publishers, Boston, **1985**

400526 Molecular Interpretation of Parameters in the Van der Waals Theory of Cubic Equations-of-State

Abstract:

Even though it is universally known that Van der Waals 1873 theory of cubic equations of state predicts inaccurate liquid-phase properties, unreliable for highly polar and hydrogen-bonded fluids, inappropriate for fluid critical point, unsuitable for large components and incapable of accurate virial coefficients, but as yet it is not widely stated that those fluid properties depend on the molecular structure of pure substances which is *evidently* not reflected by the parameters in the VDW theory. This poster suggested the possibility of including two inspired parameters into the Van der Waals 1873 equation of state. The physically-based empirical parameters that replicate molecular shapes and molecular structure of pure substances are introduced into the reformed Van der Waals (VDW) 1873 equation to reconcile the VDW 1873 theory with the stipulated shapes of the potential energy and length scale of the statistical mechanical potential functions.

By relating two of the empirical parameters (α, β) in the reformed VDW 1873 equation (which has been affectionately named the Lawal-Lake-Silberberg (LLS) cubic equation) to the microscopic force-parameters derived from the statistical mechanics energy potential functions and performing perturbation and other sensitivity analysis; the *sensitivity* of the derivatives of the structural parameters (P_c, Z_c, T_c) and the *sensitivity* of the derivatives of second, third and fourth virial coefficients with respect to α show non-linear trend with α-parameter and thus, α replicates *structure* of pure substances in the configuration of the LLS equation. Similar analysis of the derivatives of the critical volume (V_c) and Boyle temperature (T_B) with respect to β show non-linear trend with β-parameter and thus β reflects *shapes* of pure substances in the configuration of the LLS equation. Nonetheless, re-constructing the VDW 1873 cubic equation from the microscopic to the macroscopic level has been very challenging because the attraction term of the 1873 theory comes from the VDW intuition, as seen in the Schematic Figure. However, those empirical parameters reflecting structure and shape of pure substances in the LLS equation resolve the major disagreement between the VDW theory and fluid properties, including fluid critical point.

Since the shape-structure parameters (α, β) occur in the intuitive-part of the VDW theory, we can speculate that instinctive reasoning of Van der Waals is *correct* but rather Van der Waals *forgot* to carry his theory of cubic equations of state to logical conclusion before passing away in 1923. We can apply that inference to attest to the widely known but not often said that "*some pertinent variables must have been overlooked in the development of cubic equations of state*," which was echoed by Sydney Young (1890) in regard to the analysis of the coexistence gas-liquid volumes and later repeated by Stanley Walas (1985) in regard to the State of The Arts Review of cubic equations of state in his book: *Phase Equilibria in Chemical Engineering*. Stanley Walas' conclusion is "it does appear, however, that experimental data often are more nearly in accord with the *principle of corresponding states* than with specific EOS," which reinforced the statement by Guggenheim (1945) that the corresponding states principle "may safely be regarded as the most useful by-product of the Van der Waals 1873 equation of state."

The Schematic Figure is, however, contradictory to the widely accepted viewpoint of molecular theory in which repulsive forces are responsible for structure of liquids while the attraction forces play no role whatsoever in the theory of liquids. In contrast, the Schematic Figure illustrates that the attractive pressure term of the VDW equation is responsible for structure of liquids. Also, the Schematic Figure only assists in the understanding of the thermodynamic properties of dense fluids from macroscopic viewpoint *not* the interpretation of those properties in terms of the forces between the molecules. The conjecture of the Schematic Figure makes no claim whatsoever to quantify geometrical shape and structure of molecules (i.e. bonding distance and angles) which are in the realm of the applications of quantum chemistry as exemplified by the works of 1998 Nobel Laureates John Pople and Walter Kohn. Furthermore, the Schematic Figure cannot apply at molecular level to compute shapes of molecules because the shapes of Potential Energy Surface resulted from interplay of *four* basic interaction energies (electrostatic, exchange, induction, and dispersion) as discussed in *Chemical Reviews*.

While the shape-structure parameters are weighted by molecular size (or molecular volume) in the Schematic Figure, it is the *discriminant* of the attractive pressure term in the VDW equation that is responsible for accurate saturated liquid densities because $Z_c = f(\alpha, \beta)$. As typified by the coefficient of the logarithm term and the expression in the ln{} of the fugacity coefficient being depended on the discriminant, as $Z_c = f(\alpha, \beta)$. Since the Z_c factor solely depends on the limiting critical volume parameter Ω_w (or b/v_c) and they both can be expressed as $f(\alpha, \beta)$, they should be specified individually in cubic equations of state. In that regard, however, we may have unintentionally addressed the lament of Van der Waals which he stated in the 1910 Nobel Prize Lecture[1-2] as "*And here I have come to the weak point in the study of the equation of state. I still wonder whether there is a better way. In fact this question continually obsesses me, I can never free myself from it, and it is with me even in my dreams*." A better way is to use Z_c and Ω_w to specify all the parameters of cubic equations of state because Z_c solely depends on Ω_w and that would fulfil the VDW stated objective[2]: "*to determine the relation between p, v and T for a substance.*"

400545 Rationale for the Parameterization of Van Der Waals Equations-of-State: The Lawal-Lake-Silberberg Equation As Classic Prototype

Abstract:

As of yet and by judging from the State-of-The-Arts Reviews by Leland (1968), Tsonopoulos-Prausnitz (1969), Vera-Prausnitz (1972), Abbott (1973), Storvick-Sandler (1977), Horvath (1974), Prausnitz (1977), Firoozabadi (1978), Martins (1979), Wong-Prausnitz (1985), Walas (1985), Anderko (1990), Wu-Prausnitz (1998), Poling et al. (2001), Wang-Pope (2002), Valderrama (2003) and Prausnitz -Tavares (2004), the issue of which Van der Waals (VDW) model with the same number of parameters is best does not have a universal and widely acceptable answer and there is still no obvious choice among the three- and four-parameter cubic equation of state models.

This contribution asserts the reformed four-parameter VDW 1873 equation of state, which has been affectionately named the Lawal-Lake-Silberberg (LLS) cubic equation, as the preferred choice among all categories of the VDW type of cubic equations of state by honoring the *four* properties of the VDW theory, the *four* critical constraint criteria and by being in accord with the theory of cubic polynomial equations of state which stipulates four-parameter for all cubic polynomial equations. A theoretical justification within the framework of the VDW empiricisms is provided for the LLS equation as the Substance-Based Cubic Equation of State for the individual pure substances, in agreement with the VDW objective which he stated toward the end of the 1910 Nobel Prize Lecture*** as "*to determine the relation between p, v and T for a substance*." However, we are still lacking the microscopic-level description for the two empirical parameters introduced into the reformed VDW 1873 equation, the LLS equation.

The LLS equation is applicable to all categories of pure substances, regardless of the chemical structure, polarity or molecular size and prediction of accurate volumetric and thermodynamic properties over the entire PVT states are guaranteed. But, the extension of the combining mixture rules to ternary and multicomponent systems is still being explored.

*** Van der Waals, J. D., **1910** Nobel Prize Lecture available from "J. D. van der Waals - Nobel Lecture: The Equation of State for Gases and Liquids". http://www.nobelprize.org/nobel_prizes/physics/laureates/1910/waals-lecture.html

400871 Application of Artificial Neural Networks to Modelling Vapour–Liquid Equilibrium in the Fischer–Tropsch Reaction

Abstract:

Modelling vapour–liquid equilibrium (VLE) requires a choice of the equation of state (EOS), the choice of pure component parameters, and the choice of mixing rules. The more constants the EOS has, the more mixing rules are required, and the more data is required for evaluating pure-component parameters. The Fischer–Tropsch (FT) reaction mechanism has been studied for over 90 years yet an adequate description of its kinetics still seems to escape us. Including VLE in modelling an FT reactor only compounds the intricacy.

The field of computational intelligence is revolutionizing many areas of science and engineering. The ability to extract useful insights from vast amounts of weak and incomplete information is not only fuelling the current interest in "big data", but has also enabled rapid progress in more traditional disciplines such as computer vision, estimation, and robotics, where data are available but difficult to interpret.

The FT synthesis face similar challenges, as many molecular and reaction pathways are still poorly characterized and available data are incoherent. Therefore, machine learning techniques such as artificial neural networks may be instrumental in overcoming these challenges by modelling the dependencies between random variables and using them to extract and accumulate the small amounts of information each random event provides. Hence, this study applies artificial neural networks to the complex problem of modelling VLE in the FT reactor. The data used to train, validate, and test the model is obtained from literature.

15AIChE
Spring Meeting, Austin, TX

Role of Surface Chemistry in Predicting Viscosity of Concentrated Carbonaceous Solid-Water Slurries

Amrita Mukherjee, Sarma V. Pisupati#
John and Willie Leone Department of Energy and Mineral Engineering and the EMS Energy Institute,
110 Hosler Building,
The Pennsylvania State University, University Park PA 16802
#corresponding author: sxp17@psu.edu

Prepared for Presentation at
American Institute of Chemical Engineers
2015 Spring Meeting
11th Global Congress on Process Safety
Austin, Texas
April 27-30, 2015

Role of Surface Chemistry in Predicting Viscosity of Concentrated Carbonaceous Solid-Water Slurries

Amrita Mukherjee, Sarma V. Pisupati#
John and Willie Leone Department of Energy and Mineral Engineering and the
EMS Energy Institute,
110 Hosler Building,
The Pennsylvania State University, University Park PA 16802
#corresponding author: sxp17@psu.edu

Keywords: viscosity, coal, slurry, surface chemistry, inter-particle interaction

Abstract

Viscosity prediction of carbonaceous solid-water slurries is essential to process design and control. Existing viscosity models take into account particle-size distributions but fail to account for the surface chemistry of the solids. Surface chemistry affects inter-particle and particle-water interactions and therefore influences slurry viscosity of concentrated suspensions. On the basis of surface chemistries characterized by contact angle and zeta potential measurements, this study determines the inter-particle interaction energies. Polar interaction energy, observed to be 2-3 orders of magnitude greater than the electrostatic interaction energy and the van der Waal interaction energy, is clearly the dominant interaction energy for such a system. Both hydrophobic and hydrophilic interactions of these particles in water result in microstructures which trap water either in the form of coalescing droplets in aggregation networks or in the form of hydration layers around carbonaceous solids, leading to an increase in the effective solid volume fraction and thus increase in slurry viscosity. The increase in effective solid volume fraction is dependent on the surface chemistry of the solid and thus is specific to a carbonaceous solid. The factor by which solid volume fraction increases for a particular carbonaceous solid was determined using viscosity measurements and Krieger-Dougherty equation and was found to correlate very well with the oxygen to carbon ratio of the solid. Incorporating this factor in the Krieger-Dougherty equation resulted in accurate prediction of slurry viscosity. This modified model was successfully validated using three other concentrated carbonaceous solid-water slurries.

1. Introduction

Carbonaceous solid-water slurries are concentrated suspensions of coal or petcoke particles in water and are used as gasifier feedstock [1][2]. A very high solids loading, in the range of 60-75% (by weight), results in an increased slurry viscosity [3]. The slurry viscosity determines the fluid flow behavior and the pumping energy requirement. Accurately determining slurry viscosity is a challenging task; but it is also vitally important. Solids loading in the slurries should be high for gasification efficiency but at the same time viscosity of the slurries should not be too high. Higher viscosity leads to difficulties in handling and higher pumping energy consumption. The challenge in determining slurry viscosity is complicated primarily due to variables such as different particle-size distributions and widely varying surface chemistry of the carbonaceous solids along with the additives used and pH of the slurry.

Existing models for predicting viscosities of suspensions are mainly semi-empirical equations predicting viscosity on the basis of solid volume fraction and particle-size distribution [5][6]. Since the models were originally developed for non-interacting hard spheres, these do not take into account surface chemistry of the solids [7]. Among these semi-empirical equations, Krieger-Dougherty equation was proved to be the most effective semi-empirical model and therefore has been used in this work [8][9][10]. Other models such as the phenomenological model developed by Dooher for viscosity prediction accounted for surface chemistry through statistical correlations [11]. Moreover their work only involved bituminous and sub-bituminous coal and did not include the entire range of carbonaceous solids. Usui proposed a thixotropic viscosity model which can take into account the agglomerative nature of the hydrophobic carbonaceous solids but is not applicable for those carbonaceous solids like low rank coals where slurry viscosity is affected due to adsorption or absorption of free water and not aggregation of solids [12][13]. Therefore, clearly there is need for a viscosity model which can take into account surface chemistries of the entire range of carbonaceous solids, along with particle-size distribution, to predict viscosity accurately.

Surface chemistry of coals depend on the rank of the coal. Low rank coals have higher oxygen, moisture and mineral matter content as compared to high rank coals, petcoke and bitumen. For the same particle-size distribution and same solids loading, these carbonaceous solid-water slurries have widely differing viscosities owing to the differences in their surface chemistries[14].

This study follows a fundamental approach to analyze the effect of surface chemistry on slurry viscosity. Different surface chemistries lead to different particle-particle and particle-medium interactions which in turn result in different suspension microstructures. These suspension microstructures cannot be studied using the traditional, highly sophisticated light based techniques because of the micron or even millimeter sized particles of solids in the slurries and the high concentration of the slurries. Therefore in order to determine the suspension microstructure, this study calculates the inter-particle interaction energies on the basis of contact angle and zeta potential measurements of the carbonaceous solids. On the basis of the understanding of the inter-particle interaction mechanism and thus the suspension microstructure, the semi-empirical Krieger-Dougherty equation is modified to account for the surface chemistry of the carbonaceous solid. The modified equation is validated using three other carbonaceous solid-water slurry samples and provides an accurate prediction of viscosity.

2. Experimental

2.1. Materials. Pust coal (lignite), Dietz coal (subbituminous), Illinois #6 coal (bituminous), Orchard coal (anthracite) and petcoke were ground in a laboratory-sized ball mill and were then screened and separated into the following size ranges: -250+212 μm, -212+180 μm, -180+150 μm, -150+125 μm, -125+106 μm, -106+90 μm, -90+53 μm, -53+44 μm, -44+20 μm, -20 μm. Experiments were carried out with slurries of these carbonaceous solids in water. The same particle-size distribution, as shown in figure 1(a) was used for all the experiments excepting the experiments concerning model validation. Particle-size distribution was maintained constant to highlight the effect of surface chemistry of the carbonaceous solids on slurry viscosity.

Model validation was performed using petcoke, Pittsburgh no. 8 (bituminous coal) and Beulah (lignite coal) coal-water slurries. Different particle-size distributions and two different carbonaceous solid samples were used to observe the sensitivity of the model to particle-size distribution and surface chemistry. Particle-size distributions for the three samples used for model validation are shown in figure 1(b). Proximate and ultimate analyses of all the samples are presented in table 1:

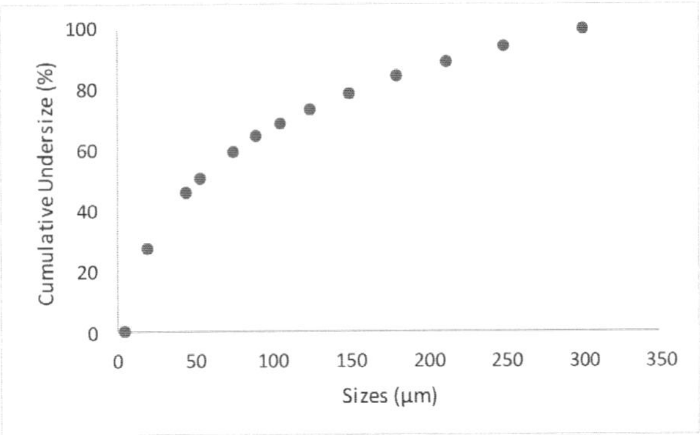

Fig 1 (a): Particle-size distribution of the carbonaceous solid particles in their respective slurries

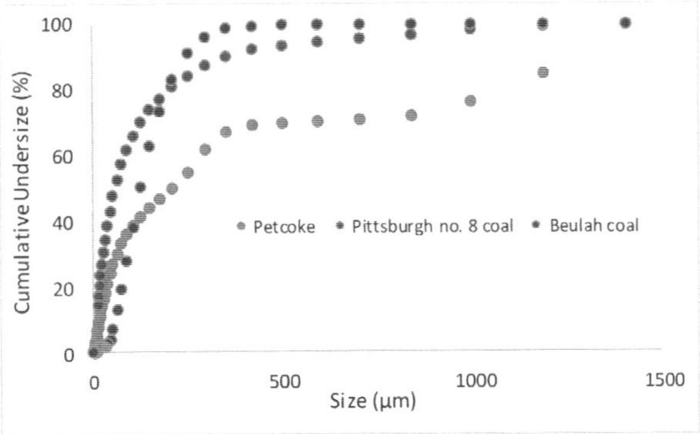

Fig 1 (b): Particle-size distribution of the carbonaceous solid particles in their respective slurries in the model validation experiments

Table 1: Proximate and ultimate analyses results:

	Proximate Analysis (weight %, as received basis)				Ultimate Analysis (weight %, as received basis)				
	M	VM	A	FC	C	H	N	S	O
Petcoke	0.32	10.96	0.17	88.55	89.08	3.64	1.47	5.43	0.20
Orchard coal	1.59	9.27	10.79	78.35	80.80	3.52	1.91	1.14	0.24
Pittsburgh no.8 coal	0.31	35.70	9.80	54.19	76.68	4.93	1.39	1.46	5.42
Illinois#6 coal	3.31	38.46	12.53	45.70	67.50	4.75	1.45	5.72	5.67
Dietz coal	8.46	41.69	3.87	45.99	70.70	4.91	0.58	0.39	11.8
Pust coal	11.12	41.17	8.55	39.16	60.77	4.08	1.28	0.60	13.60
Beulah coal	18.88	39.19	5.19	36.74	54.94	3.87	1.32	0.60	15.20

Note: M, VM, A, FC stand for moisture, volatile matter, ash, fixed carbon respectively.

2.2. Preparation of carbonaceous solid-water slurry. Calculated amount of carbonaceous solids (calculated on the basis of targeted slurry concentration) was added to weighed amount of distilled water. The mixture was then stirred by a propeller agitator for 10 minutes and viscosity was measured [1]. Solids loadings of carbonaceous solid - water slurries were calculated and reported on as received basis.

2.3. Viscosity. Viscosity was measured using Bohlin 88 Viscometer (a product of Malvern Instruments Ltd., U.K.) The shear rate used for all the measurements was $100s^{-1}$ in keeping with the shear rate used in industry. All viscosity measurements were performed in the temperature range of 23 to 26 $^{\circ}C$ (ambient temperature).

2.4. Contact Angle. Contact angles were measured using a goniometer (ramé-hart Model 295). Contact angles of water, glycerol and diiodomethane, on the carbonaceous solid surfaces were measured using the Sessile drop technique.

2.5. Zeta Potential. Zeta Potential of the carbonaceous solid samples were measured using Malvern Zetasizer ZS which uses electrophoretic light scattering technique.

3. Results and Discussions

3.1. Interaction Energy Calculations & Analyses. Van der Waal interaction energies and polar interaction energies were calculated on the basis of contact angle measurements whereas the electrostatic interaction energies were calculated using zeta potential measurements. Contact angle values were also used to determine the surface energy parameters. Detailed calculations of the inter-particle interaction energies have been included elsewhere [15]. The observations of the interfacial energy calculation can be summarized as:

1. For each of the concentrated carbonaceous solid–water slurries polar (hydrophobic/hydrophilic) interaction energy is 2 to 3 orders of magnitude greater than the van der Waal or electrostatic interaction energy. Therefore, polar interaction energies play the key role in determining slurry rheology of highly concentrated slurries.

2. For slurries of petcoke, Orchard coal, Illinois no. 6 coal and Dietz coal in water, the inter-particle polar interaction energy is attractive hydrophobic whereas for Pust coal it is repulsive hydrophilic in nature. The magnitude of attractive hydrophobic energy decreases from petcoke to Dietz coal in the order of increasing oxygen to carbon ratio.

Analyses based on the above observations are as follows:

1. Petcoke and Orchard coal particles in water form aggregation network due to the predominant hydrophobic inter-particle interaction energy. The aggregation networks trap free water within the network structure. Since the contact angles of water on petcoke and Orchard coal surface is greater than 100°, the water can only form unstable coalescing droplets on such surfaces. The lack of oxygen functional groups on the surfaces of these solids contribute to its apolarity.[16]

2. The inter-particle hydrophobic interaction energy is less in magnitude for the Illinois no. 6 and the Dietz coal which implies that the aggregation networks in the slurries would be more like loose flocs. Since water contact angle on the surfaces of Illinois no. 6 and Dietz coal are 73° and 42° respectively, water would not form coalescing droplets and actually start forming hydrogen bonds with oxygen functional group on the surface of the carbonaceous solids. Therefore free water from the slurry would start forming stable hydration layers around carbonaceous solids.

3. Formation of aggregation network structures would not be possible in Pust coal water slurries due to the existence of inter-particle hydrophilic repulsive energy. Oxygen functional groups on the surface of such carbonaceous solids form hydrogen bonds with the free water and thus trap free water to form hydration layer around the particles.

4. In all this cases, it is evident that free water is lost from the slurry either in the form of coalescing droplets or hydration layers. Loss of free water reduces fluidity of the slurry and affects the effective solid volume fraction. Solid volume fraction is given by:

$$\text{Solid Volume Fraction } (\varphi) = \frac{Volume\ of\ Solid}{Volume\ of\ Solid + Volume\ of\ free\ water} \quad (1)$$

It is evident from equation (1) that as volume of free water decreases, effective solid volume fraction increases. Now, the volume of free water lost or the volume of trapped water depends on whether the water is lost as coalescing droplets or hydration layers and also on the thickness of hydration layer. Therefore clearly a loss of free water or the increase in effective solid volume fraction, is a function of the surface chemistry of the solid.

3.2. Viscosity Prediction of Carbonaceous Solid-Water Slurries.

The increase in effective solid volume fraction of the carbonaceous solids in suspension with water was determined with the help of measured viscosity values and Krieger-Dougherty equation. The particle-size distributions of the carbonaceous solids were used to calculate maximum solid volume fraction (φ_m) using Veytsman's equation [17]. Since the particle-size distribution was maintained same for all carbonaceous solids-water slurries for this part of the experiment, the calculated φ_m value was 0.74 in all the cases. The experimentally measured viscosities were used in the Krieger-Dougherty equation (given by equation 2) to calculate the final effective solid volume fraction. The factor by which effective solid volume fraction increases or volume fraction correction factor (VFCF) was calculated for all the solids and was found to vary for different carbonaceous solids. VCFC values for each of the carbonaceous solids in water are listed in table 2. For each of these carbonaceous solid-water slurries, initial solids volume fractions were varied till a point until the slurry ceased to flow (i.e. point where the slurry had infinite viscosity).

$$Measured\ Viscosity = \eta_r = \left(1 - \frac{\varphi_f}{\varphi_m}\right)^{-[\eta]\varphi_m} \quad (2)$$

φ_f= final solids loading (dry basis): determined using equation 10

φ_i= initial solids loading (dry basis): already known

It can be observed from table 2 that the VCFC values, increase from hydrophobic to hydrophilic solids implying that the increase in effective solid volume fraction is more for hydrophilic solids. Hydrophilic solids having higher oxygen to carbon ratio can attract more water molecules and form thicker and more stable hydration layers. On the other hand the extremely hydrophobic petcoke and Orchard coal can only trap water in the form of unstable coalescing droplets. It's easier to release trapped water by applying shear in the latter case. A slight variation of VFCF values of a particular carbonaceous solid in water was observed over different solid volume fractions (as indicated by the small standard deviation). The small variation is due to the fact that as the initial solid volume fraction increases, it leads to more and more crowding of the solid particles in the water, leaving less space for the entrapped water or the hydration layer.

Figure 2 shows the variation of VFCF values as a function of the oxygen and carbon content of the carbonaceous solid

Table 2: VFCF values of the different carbonaceous solids in water

	Carbonaceous solids water slurries				
Initial Solids Volume fraction	Petcoke	Orchard coal	Illinois#6 coal	Dietz coal	Pust coal
.36	N/A	N/A	N/A	N/A	1.85
.38	N/A	N/A	N/A	N/A	1.82
.40	N/A	N/A	N/A	1.7	1.77
.42	N/A	N/A	N/A	1.65	1.74
.44	N/A	N/A	1.54	1.59	1.67
.46	N/A	N/A	1.52	1.54	Infinite Viscosity
.48	N/A	1.36	1.47	1.48	
.50	N/A	1.35	1.42	Infinite Viscosity	
.52	1.32	1.33	Infinite Viscosity		
.54	1.31	1.3			
.56	1.27	1.27			
.58	1.24	Infinite Viscosity			
Mean	**1.28**	**1.31**	**1.49**	**1.59**	**1.77**
Standard Deviation	**.03**	**.04**	**.053**	**.07**	**.07**

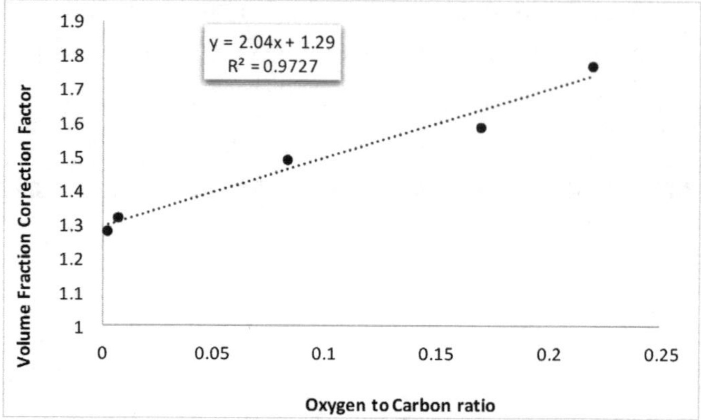

Fig 2: Variation of VFCF with oxygen to carbon ratio

VFCF was observed to vary linearly and correlate well with the oxygen to carbon ratio. Therefore for any carbonaceous solid whose oxygen and carbon content are known , their VFCF can be calculated from equation 3 and used in equation 4 to predict their corresponding slurry viscosities accurately.

$$VCFC = 2.04\frac{O}{C} + 1.29 \tag{3}$$

$$\eta_r = [1 - \frac{\left(2.04 * \frac{O}{C} + 1.29\right)\varphi}{\varphi_m}]^{-[\eta]\varphi_m} \tag{4}$$

Where,

O = Oxygen content of the carbonaceous solid on dry basis

C = Carbon content of the carbonaceous solid on dry basis.

Other terms in the equation have already been defined in the previous sections.

3.3. Validation of the modified model.

The modified model was validated using slurries of petcoke, Pittsburgh no.8 coal (bituminous coal) and Beulah coal (lignite coal) particles in water. Maximum packing efficiencies of these samples were determined on the basis of their particle-size distributions using Veytsman's model. Oxygen to carbon ratios as obtained from ultimate analyses of the samples were used in equation 2 to obtain the VFCFs of the corresponding carbonaceous solids in water. Maximum packing efficiencies (φ_m) and VFCFs, inputs to the modified Krieger-Dougherty equation, are reported in table 3.

Table 3: Input parameters to the modified Krieger-Dougherty equation

	Maximum packing efficiency	VCFC
Petcoke	.78	1.28
Pittsburgh no.8 coal	.76	1.43
Beulah coal	.86	1.86

Experimentally measured viscosities of these slurries were then compared with predictions of both Krieger-Dougherty equation and the modified Krieger-Dougherty equation as shown in figures 3 (a-c).

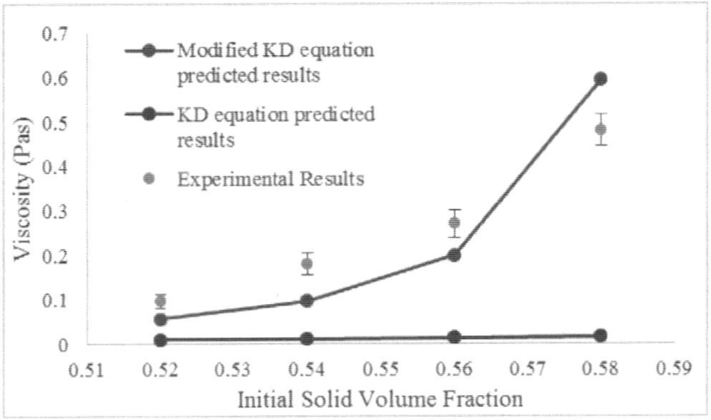

(a) Petcoke-water slurry

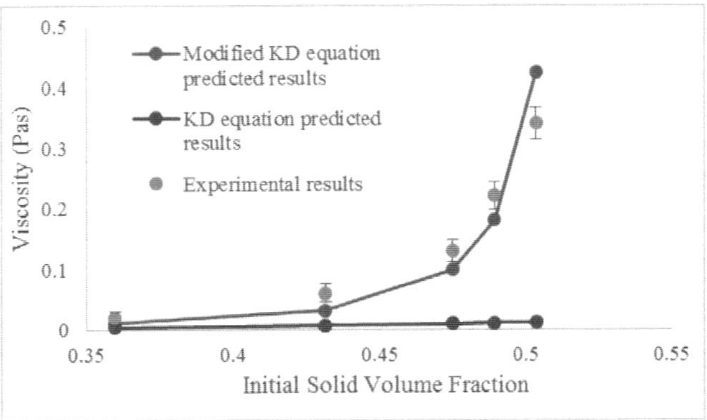

(b) Pittsburgh no.8 coal-water slurry

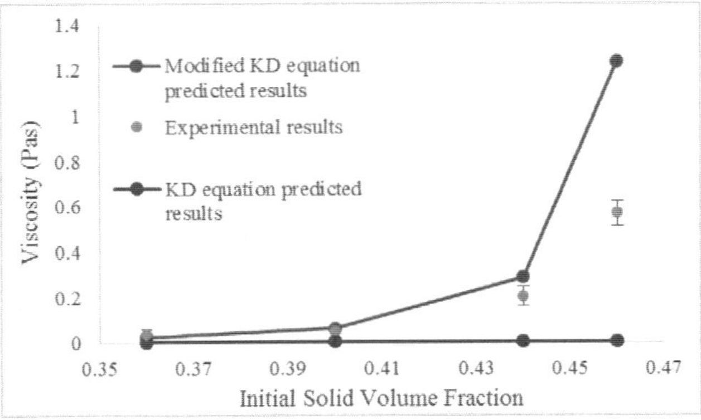

(c) Beulah coal-water slurry

Figure 3: Comparison of experimentally measured viscosities with model predicted viscosities as a function of initial solid volume fraction

It is evident from the figures that modified equation predicts viscosity much more accurately especially at higher solid loading compared to original Krieger-Dougherty equation. At higher solid loading owing to minimal inter-particle distance, inter-particle interaction plays a major role in governing slurry viscosity. The modified equation takes into account the surface interactions, and therefore its predictions are more accurate.

4. Conclusions:

1. Inter-particle polar interaction energies play a dominant role in deciding carbonaceous-solid water slurry microstructure.

2. Carbonaceous solids in water have the tendency to trap free water either in the form of coalescing droplets or in the form hydration layers. The loss of a part of the free water from the slurries reduces the fluidity of the slurry by increasing the effective solid volume fraction. The increase in solid volume fraction is characteristic of a particular carbonaceous solid in water and therefore represents surface chemistry of the carbonaceous solid.

3. Increase in solid volume fractions of carbonaceous solids in water were calculated using measured viscosity values and Krieger-Dougherty equation. The increase in solid volume fraction was found to increase with increase in oxygen functional groups present on the surface of the solids

4. VCFC (increase in effective solid volume fraction) was observed to correlate well with the oxygen to carbon ratio of the carbonaceous solid. Therefore for a carbonaceous solid with known oxygen to carbon ratio, its VCFC can be predicted from the correlation and by using the VCFC value in the modified Krieger-Dougherty equation its viscosity can be predicted accurately.

5. Modified model was validated with three other sample slurries and was found to make accurate prediction of slurry viscosity.

5. Future Work:

- Volume of loss of free water for each of carbonaceous solid water slurries is to be confirmed by Differential Scanning Calorimetry.
- This model would be further developed to take into account effect of additives and slurry pH on slurry viscosity.
- Expand the applicability of the model so as to predict viscosities of even non-carbonaceous solid-water slurries such as proppant water slurry.

6. References:

[1] N.-S. Roh, D.-H. Shin, D.-C. Kim, and J.-D. Kim, “Rheological behaviour of coal-water mixtures. 1. Effects of coal type, loading and particle size,” *Fuel*, vol. 74, no. 8, pp. 1220–1225, Aug. 1995.

[2] F. Gao, J. Liu, C. Wang, J. Zhou, and K. Cen, “Effects of the physical and chemical properties of petroleum coke on its slurryability,” *Petroleum Science*, vol. 9, no. 2, pp. 251–256, Jul. 2012.

[3] H. Dinçer, F. Boylu, a. . Sirkeci, and G. Ateşok, “The effect of chemicals on the viscosity and stability of coal water slurries,” *International Journal of Mineral Processing*, vol. 70, no. 1–4, pp. 41–51, Jun. 2003.

[4] S. Mueller, E. W. Llewellin, and H. M. Mader, “The rheology of suspensions of solid particles,” *Proceedings of the Royal Society of London A: Mathematical, Physical and Engineering Sciences*, vol. 466, no. 2116, pp. 1201–1228, Feb. 2010.

[5] M. Mooney, “The viscosity of a concentrated suspension of spherical particles,” *Journal of Colloid Science*, vol. 6, no. 2, pp. 162–170, Apr. 1951.

[6] R. Simha, “A Treatment of the Viscosity of Concentrated Suspensions,” *Journal of Applied Physics*, vol. 23, no. 9, p. 1020, 1952.

[7] A. Mukherjee, P. Rozelle, and S. V Pisupati, “Influence of Surface Properties and Applicability of Current Viscosity Models for Carbonaceous Solid Water Slurries,” accepted.

[8] T. Tadros, “Interparticle interactions in concentrated suspensions and their bulk (rheological) properties.,” *Advances in colloid and interface science*, vol. 168, no. 1–2, pp. 263–77, Oct. 2011.

[9] I. M. Krieger, “A Mechanism for Non-Newtonian Flow in Suspensions of Rigid Spheres,” *Journal of Rheology*, vol. 3, no. 1, p. 137, 1959.

[10] W. Pabst, “Fundamental Considerations On Suspension Rheology,” *Ceramics-Silikaty*, vol. 48, pp. 6–13, 2004.

[11] J. P. Dooher, N. Malicki, and J. Trudden, *Physiochemical principles of coal-water-slurry gasifier feedstocks*. 1990, p. 151.

[12] H. Usui, “A thixotropy model for coal-water mixtures,” vol. 60, 1995.

[13] H. Usui, "Viscosity Prediction of Dense Slurries with Agglomerative Coal Particles.," *Nihon Reoroji Gakkaishi*, vol. 29, no. 2, pp. 71–76, 2001.

[14] M. Pawlik, J. S. Laskowski, and H. Liu, "Effect of Humic Acids and Coal Surface Properties on Rheology of Coal-Water Slurries," *Coal Preparation*, vol. 18, no. 3–4, pp. 129–149, Jan. 1997.

[15] A. Mukherjee and S. V Pisupati, "Inter-particle interactions in highly concentrated coal-water slurries and their effect on slurry viscosity," *Energy & Fuels*, submitted.

[16] G. D. Botsaris, *Interfacial Phenomena in Coal Technology*. 1988, p. 464.

[17] B. Veytsman, J. Morrison, A. Scaroni, and P. Painter, "Packing and Viscosity of Concentrated Polydisperse," *Energy & Fuels*, vol. 0624, no. 6, pp. 1031–1039, 1998.

401697 Optimizing Crude Oil Separations – a Study of Turbulence and Chemical Composition Influence on the Drop Size Distribution of Crude Oil in Maritime Conditions (Electronic Poster Presentation)

Abstract:

Optimizing crude oil separations – a study of turbulence and chemical composition influence on the drop size distribution of crude oil in maritime conditions

Efficiency in the exploration of crude oil deposits is of great importance for the oil and gas industry. With the usage of large amounts of feed water in the process comes the need for fast and efficient separation processes for the liquid/liquid dispersion. The quality of the separation translates directly into costs for storage, transportation and subsequent processing. The dispersion is thermodynamically unstable and given time, the droplets will separate out to form two bulk liquid phases. The drop size distribution is the most relevant parameter for this type of separation process. Great efforts have been made to measure the size of liquid dispersed droplets. The utilized techniques can be divided into those that require samples to be withdrawn from the vessel or pipe, and those that can be applied in situ.

For deep sea analysis in-situ methods are preferential since they account of the significant impact of the present conditions on the size of the drops. Many investigations of droplet size distributions or a representable diameter like d_{32} have focused on highly dilute dispersions, where it is assumed that coalescence can be eliminated. As this approach provides results with limited practical application, a measurement technology for high concentrations is necessary. In comparison to chemical, pharmaceutical, and food manufacturing, the petroleum industry presents additional challenges, as the optical conditions of oil streams are much more complex then specialty and fine chemicals. Also the variation of fluid physical properties and process conditions is stronger. The understanding, prediction and control of the drop size in crude oil applications is rather challenging. A detailed description of the opposing phenomena, drop breakage and coalescence, is paramount for successful mathematical modeling. An additional difficulty in handling such dispersed systems is the probability of phase inversions (PI). In case of a spontaneous occurrence within a process, the consequences for product and equipment can be catastrophic [1]. On the contrary, it can be used for a more controlled handling of the described applications. Therefore a better understanding and modeling of this phenomenon is desirable.

This study reports on a turbulent liquid/liquid system using original North Sea crude oils and sea water was investigated. The operation parameters dispersed phase fraction, turbulence intensity and surfactant concentration have been varied systematically. The transient drop sizes as the unstable behavior of the water and organic phase have been observed using an in-situ microscope [2]. The drop size was measured using a fully automated image analysis tool [3]. These data give a reliable base to understand liquid-liquid separations better.

[1] Nienow, A. W., Adv. Colloid Interface Sci. 2004, 108, 95-103.

[2] Maaß, S., Wollny, S., Voigt, A., Kraume, M., Exp. Fluids 2011, 50, 259-269.

[3] Maaß, S., Rojahn, J., Hänsch, R., Kraume, M., Comput. Chem. Eng. 2012, 45: 27-37.

401743 Prediction of High-Temperature and High-Pressure Dynamic Viscosity of Hydrocarbons By Van Der Waals Transport Equation of State

Abstract:

According to the review of liquid viscosity by Brush (1962), [1] "there is no general agreement on whether viscosity is essentially due to the attractive and repulsive forces," but on macroscopic scale both viscosity and density are function of the thermodynamic equilibrium state of the fluid[2] (therefore, they both surrender to the Law of Corresponding States [3,4]) while on microscopic scale both properties reflect the effects of molecular motion and interaction. So, in accordance with the hypothesis enunciated by Phillips (1927)[5] and later improvements by Little-Kennedy (1966)[6] and Lawal (1986), [7] a method is established for predicting Newtonian viscosity or dynamic viscosity (η) over the entire PVT states of single-phase fluids such as polar and nonpolar substances, hydrocarbon, non-hydrocarbon, petroleum fractions and high-molecular-weight-fluids of the type currently being measured for high-temperature and high-pressure (HTHP) deep-water formations [8-17] and carbon dioxide sequestration.

By using the1986 formalism of the Viscosity Equation of State (VEOS) development, the VEOS technique is designed on the basis of the phenomenological similarity between PVT and TηP spinodal graphs and is solely depended on accurate critical viscosity of pure substances. There is no resorting to the judicious choice of liquid density correlation or any volumetric properties because the coexistence gas-liquid viscosities, the dilute-gas viscosity, the dense and supercritical fluid viscosities are based on the four-parameter cubic equation of state[7] which is valid over the entire PVT states, include the vapor-liquid critical point.

The VEOS is validated by the prediction of accurate coexistence gas-liquid viscosities over the entire PVT states of pure substances, including polar and nonpolar substances, inert gases (Argon, Helium, Xenon), hydrocarbons of high-molecular-weights and non-hydrocarbons (nitrogen, carbon dioxide, hydrogen sulfide) and liquid viscosity of very viscous fluids.

References:

1. Brush, S. G., "Theories of Liquid Viscosity," *Chem. Rev. 62*, 513, **1962**

2. Brule, M. R., Starling, K. E., *Ind. Eng. Chem. Process Des. Dev*., *23*, 833, **1984**

3. Ely, J. F., Hanley, H. J. M., "Prediction of Transport Properties: 1. Viscosity of Fluids and Mixtures," *Ind. Eng. Chem. Fund. 20*, 323-332, **1981**

4. Ely, J. F., Hanley, H. J. M., "Prediction of Transport Properties: 2. Thermal Conductivity of Pure Fluids and Mixtures," *Ind. Eng. Chem. Fund. 22*, 90-97, **1983**

5. Phillips, P., "The Viscosity of Carbon Dioxide," *Proc. Roy. Soc. 87A*, 48, **1927**

6. Little, J. E., Kennedy, H. T., "A Correlation of the Viscosity of Hydrocarbon Systems with Pressure, Temperature and Composition," *Soc. Pet. Eng. J*., 157 (June**1968**)

7. Lawal, A. S., "Prediction of Vapor and Liquid Viscosities From the Lawal-Lake-Silberberg Equation of State," In *SPE/DOE Paper No.14926 in the Proceeding of SPE/DOE Fifth Symposium on Enhanced Oil Recovery of the Society of Petroleum Engineers and the Department of Energy* held in Tulsa, OK, April 20-23, **1986**

8. Caudwell, D. R.; Vesovic, V.; Trusler, J. P. M.; Wakeham, W. A., "The viscosity and Density of n-Dodecane and n-Octadecane at Pressures up to 200 MPa and Temperatures up to 473K," *Int. J. Thermophysics*, *25*, 1340-52, **2004**

9. Caudwell, D. R.; Trusler, J. P. M.; Vesovic, V.; Wakeham, W. A., "Viscosity and Density of Five Hydrocarbon Liquids at Pressures up to 200 MPa and Temperatures up to 473 K," *J. Chem. Eng. Data*, **54**, 359-366, **2009**

10. Y.L. Sen, Kiran, E., "High-pressure viscosity and density of n-alkanes," *Int. J. of Thermophysics, 13,* 411-442, **1992**

11. Burgess, W. A., Tapriyal, D., Gamwo, I. K., Morreale, B. D., McHugh, M. A., Enick, R. M., "Viscosity Models Based on the Free Volume and Frictional Theories for Systems at Pressures to 276 MPa and Temperatures to 533 K," *Ind. Eng. Chem. Res*., *51*, 16721-33, **2012**

12. Liu, K.; Wu, Y.; McHugh, M. A.; Baled, H.; Enick, R. M.; Morreale, B. D., "Equation of State Modeling of High-Pressure, High-Temperature Hydrocarbon Density Data," *J. Supercrit. Fluids*, *55*, 701, **2010**

13. Baled, H. O., Xing, D., Katz, H., Tapriyal, D., Gamwoa, I. K., Soong, Y., Bamgbade, B. A., Wu, Y., Liu, K., McHugh, M. A., Enick, R. M., "Viscosity of n-Hexadecane, n-Octadecane and n-Eicosane at pressures up to 243 MPa and temperatures up to 534 K," *J. Chem. Thermodynamics*, *72*, 108-116, **2014**

14. Ducoulombier, D.; Zhou, H.; Boned, C.; Peyrelasse, J.; Saint-Guirons, H.; Xans, P., "Pressure (1−1000 bar) and Temperature (20−100 °C) Dependence of the Viscosity of Liquid Hydrocarbons," *J. Phys. Chem.*, *90*, 1692-1700, **1986**

15. Hogenboom, D. L.; Webb, W.; Dixon, J. A., "Viscosity of Several Liquid Hydrocarbons as a Function of Temperature, Pressure, and Free Volume," *J. Chem. Phys*., *46*, 2586-98, **1967**

16. Tanaka, Y.; Hosokawa, H.; Kubota, H.; Makita, T., "Viscosity and Density of Binary Mixtures of Cyclohexane with n-Octane, n-Dodecane, and n-Hexadecane Under High Pressures," *Int. J. Thermophysics*, *12*, 245-264, **1991**

17. Assael, M. J.; Papadaki, M., "Measurements of the Viscosity of n-Heptane, n-Nonane, and n-Undecane at Pressures up to 70 MPa," *Int. J. Thermophysics*, 12, 801−810, **1991**

401790 Inversion of the Van Der Waals Parameters for Estimation of Binary Interaction Constants

Abstract:

As yet, it is still impossible to invert the equations of statistical mechanics to explicitly yield intermolecular forces as functions of the macroscopic properties of individual pure substances. But, the critical point of the van der Waals (VDW) gas-liquid transition provides a source of deriving interaction constants from the measured critical temperatures and pressures of binary systems. By inverting the three-dimensionless criticality parameters established by the VDW theory, we back-calculate the binary interaction constants for the attractive parameter and two other EOS parameters whose binary interaction constants relate mixture critical volume and compressibility factor of VDW equation to composition. The Lawal-Lake-Silberberg (LLS) equation of state (EOS) provides a prototype procedure for VDW type-equations in establishing binary interaction constants from measured critical loci of binary systems and the technique is validated by the prediction of accurate critical properties, high-pressure volumetric properties and binary high-pressure phase equilibria. It is revealed that the derived binary interaction constants are *totally* independent of temperature, pressure and composition. However, the generalization of the binary interaction constants in terms of the macroscopic properties of pure substances is still very challenging but may not be impossible through the realms of statistical mechanics.

399289 Gas Evolution Rates in Crude Oil for Upstream Applications: A Literature Review

Abstract:

As gas-liquid separators are reduced in size to accommodate offshore and subsea specifications, ensuring that the units are appropriately sized has become increasingly important. Failure to size separators correctly will hinder both the safety and reliability of the unit, and could compromise processes downstream. While the vessel volume is typically calculated from estimated flow rates and well head conditions, the rate of gas evolution during flow and at high pressures has been largely overlooked. This presentation provides an overview of the research that is relevant to gas evolution with specific emphasis on the impact of flow on the rate of gas evolution. Sources from literature show that the rate of gas evolution can be affected by parameters including initial saturation pressure, liquid temperature, rate of agitation, and the presence of solid particulates. Equilibrium and dynamic modeling is also discussed in the presentation.

15 AIChE
Spring Meeting, Austin, TX

FLUID-STRUCTURE INTERACTION MODELING OF SUBSEA JUMPER PIPE

Mohammad A. Elyyan
ANSYS, Inc.
Houston, TX USA
mohammad.elyyan@ansys.com

Yeong-Yan Perng
ANSYS, Inc.
Austin, TX USA
yeong-yan.perng@ansys.com

Ali Najafi
ANSYS, Inc.
Austin, TX USA
ali.najaf@ansys.com

Prepared for Presentation at
American Institute of Chemical Engineers
2015 Spring Meeting
11th Global Congress on Process Safety
Austin, Texas
April 27-30, 2015

UNPUBLISHED

FLUID-STRUCTURE INTERACTION MODELING OF SUBSEA JUMPER PIPE

Mohammad A. Elyyan
ANSYS, Inc.
Houston, TX USA

Yeong-Yan Perng
ANSYS, Inc.
Austin, TX USA

Ali Najafi
ANSYS, Inc.
Houston, TX USA

ABSTRACT

Flow-induced vibration (FIV) and vortex-induced vibration (VIV) are main causes of subsea piping failure, where subsea pipes experience large fluctuating forces due to internal multiphase flows and external vortex shedding. These fluctuating forces can induce severe vibrations leading to premature piping failure. This paper presents a transient numerical study of a typical subsea M-shape jumper pipe that is submerged in seawater and carrying a gas-liquid multiphase flow subject to a slug frequency of 4.4 Hz, starting from rest to include the start-up effect as part of the study. 3-D numerical simulations were used to capture the fluid-structure interaction (FSI) and estimate pipe deformations due to fluctuating hydrodynamic forces. Numerical analysis showed strong correlation between the multiphase slugging frequency and pipe deformation. Moreover, results showed startup regime can be severe with maximum deformation of 0.08 m, while maximum deformation at steady operation is close to 1 mm.

INTRODUCTION

Subsea pipes are designed to carry production fluid from well head to topside processing facilities. Production fluid consists of different phases in general, e.g. oils, gas, and/or water. Due to the density difference between phases, multiphase flows are transient exert fluctuating forces on the internal pipe. These fluctuating forces can lead to what is known as flow-induced vibration (FIV) of the pipe. On the other hand, subsea pipes are submerged in deep water and are subject to sea current. As the current flows around the jumper pipe, vortices are periodically shed from the rear side of the pipe. These shed vortices exert oscillating forces on the subsea pipe, which may lead what is known as vortex-induced vibration (VIV).

FIV and VIV can interact and have a strong effect on the structure integrity; can cause resonance and fatigue, which may lead to structure failure, losses in production and devastating environmental effects. Actually, Swindell [1] reported that up to 21% of topside pipework failures are caused by vibration-induced fatigue. Thus, it is important to accurately predict FIV, VIV and their effect on structure integrity for subsea piping transporting multiphase flows.
Several studies, both numerical and experimental, have been conducted to analyze the FIV problem in piping. Yih and Griffith [2] reported that for a multiphase flow in pipe, the slug flow has the highest momentum flux, i.e. slug flow has the largest fluctuating forces acting on the pipe walls. Moreover, Sim et al. [3] reported that for slug flow, the void fraction has the largest effect

on the RMS of the unsteady forces. Riverin et al. [4] in their study of effect of fluctuating forces on flow turning elements also reported that the fluctuating forces correlated with the void fraction of the flow.

Oil and gas production pipelines are more exposed to multiphase flows with large void fractions, which could results in large fluctuating forces acting on the pipe walls. These forces will lead to FIV, as discussed earlier. In order to address this risk, API RP 1111 [5] recommends that any pipeline design should consider dynamic loads and resulting stresses, which may be induced by vibrations due to current-induced vortex shedding and other hydrodynamic loading. Nevertheless, it does not outline the best practice on how to account for these dynamic loads.

Recently, few attempts have been made to address flow induced vibration in subsea installations. Jia [6] investigated FIV in different components of subsea equipment. His Study showed that FSI analysis of subsea piping carrying multiphase flow is important. No fatigue or modal analysis was provided in that study. Jia [7] further studied several factors affecting slug-induced vibrations in a straight pipeline span.

Lu et al. [8] numerically modeled fluid-structure interaction (FSI) of a subsea jumper pipe carrying internal flow and exposed to external sea current. They did a detailed analysis to obtain realistic operating conditions at well head. Nevertheless, no multiphase flow slugging was applied.

Elyyan et al. [9] studied FSI modeling methodology of a jumper pipe carrying slugging multiphase flow. Elyyan et al. [9] reported that it was necessary to include two-way FSI coupling to accurately capture structure deformation in in the startup region. Their study did not include external sea water damping effects.

This paper is a continuation of the previous effort to provide a detailed numerical modeling methodology for analyzing the dynamic loads acting on subsea jumper pipes. The effect of external sea current is added in this paper. This paper presents a numerical analysis of the transient hydrodynamic forces acting on an M-shape jumper pipe. The hydrodynamic forces are due to slugging multiphase flow entering the jumper, and external sea current. The internal multiphase flow is started from rest to include start-up effects.

OBJECTIVE

The main objectives of this paper are: i) to investigate the flow structure interaction of a jumper pipe carrying multiphase flows and exposed to external sea current; ii) to investigate the effect of external seawater structure deformation; iii) investigate the start-up effect on the jumper pipe structure.

COMPUTATIONAL MODEL AND BOUNDARY CONDITIONS

A typical M-shape jumper pipe geometry is used in this study, Figure 1. The pipe material was assumed to be structure steel with 2.5-inch thickness and D_i = 5.5 inches.

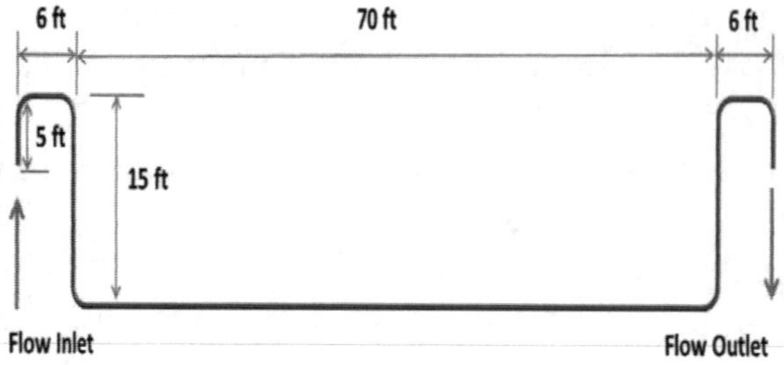

Figure 1: M-Shape Jumper Pipe Geometry

The inlet flow entering the jumper pipe consists of oil and methane gas, which is typical for production conditions. In order to simulate the worst case scenario, slugging flow conditions are imposed the pipe inlet with gas void fraction of 0.45 and slugging frequency of 4.4 Hz (f_{slug}). The jumper pipe is assumed submerged in seawater at 2000 m depth with $P_{seawater}$ = 3000 psi. Moreover, oil and methane gas are assumed to be operating at an average pressure of 2100 psi. Constant fluid properties were used for oil and ideal gas is used for Methane gas. Both ends of the jumper pipe are clamped, i.e. fixed support boundary condition.

In order to model start-up conditions, the jumper pipe is assumed to be at rest, i.e. U_{fluid}=0.0 m/s at t=0.0. Moreover, the pipe is assumed to be filled with oil initially. Start-up conditions are common in oil and gas applications, and can introduce large hydrodynamic forces on the piping system. The external sea current is assumed to be 0.4 m/s. This is the average velocity at the depth of consideration [8].

NUMERICAL METHODOLOGY

In this paper, we solve a fully coupled transient fluid-structure interaction (FSI) model for the jumper pipe carrying multiphase flow and subject to external sea current. The hydrodynamic forces acting on the wall, internal and external, are solved using CFD. These forces are then passed to the structural solver to solve for the pipe deformations. Once the deformation are estimated using the structural solver, they are passed back to update the geometry in the CFD solver. This data exchange between the two solvers repeats at every time step as the solution progresses.

In the current study, ANSYS Fluent is employed to solve for the hydrodynamic forces acting on the pipe walls. The internal multiphase flow is solved using the volume of fluid (VOF) model. In the VOF model, s a single set of momentum (Navier-Stokes) equations is solved, while the interface between phases is tracked by solving a volume fraction equation for each phase. The VOF model is suitable for immiscible fluids applications. Also, Pipe deformations and stresses are obtained by solving the stress-strain equation using ANSYS Mechanical. Moreover, coupling between the CFD and FEA solvers is managed using ANSYS System Coupling inside ANSYS Workbench.

CFD Settings: The governing equations, i.e. continuity and Navier-Stokes equations were solved using the pressure based segregated solver with the volume of fluid model (VOF) for the multiphase flow. The Realizable k-ε RANS turbulence model was used to capture the expected adverse pressure and flow separation in the pipe. The mesh was constructed using 370,000 structured hexahedral cells.

Structural Settings: Fixed support was used on both ends of the jumper pipe. The hydrodynamic pressure of the external seawater was applied to the external pipe wall. The model was meshed using 97,000 nodes with structured hexahedral mesh.

RESULTS AND DISCUSSION

Modal Analysis: To evaluate the resonant risk from FIV, a modal analysis of the jumper pipe was performed to compute its mode shapes and natural frequencies. The presence of the internal and external fluids can significantly change the vibration characteristics of the structural steel jumper pipe. To determine the extent of this effect, we have modeled the fluid-structure coupling that represents the interaction between these two domains.

To evaluate the damping effect of water on the structural steel pipe, four modal analyses have been performed to compute mode shapes and natural frequencies of the pipe containing water and submerged in water: i) jumper pipe with no additional fluid mass; ii) jumper pipe with internal damping fluid only, iii) jumper pipe with external damping fluid only; iv) jumper pipe with both internal and external damping fluid.

Table 1 lists the dominant frequencies for the different jumper pipe with(out) damping fluid cases. As can be seen from the table, the addition of the damping fluid resulted in lowering the natural frequencies of the structure. The inclusion of the internal and external damping fluids resulted in reducing the natural frequencies by 15-30% of that of the jumper structure alone.

Figure 2 shows the first four mode shapes of the jumper pipe with the case with internal and external fluid damping. As expected, the natural frequencies are mainly in plane normal to the pipe horizontal span.

Table 1: Jumper pipe mode shapes and natural frequencies.

Mode	No Damping Fluid	With Internal Fluid	With External Fluid	With External and Internal Fluid
	[Hz]	[Hz]	[Hz]	[Hz]
1	0.998	0.945	0.896	0.857
2	2.036	1.929	1.827	1.749
3	2.212	2.094	2.141	2.063
4	2.402	2.276	2.156	2.069
5	3.701	3.506	3.323	3.180
6	5.933	5.608	3.599	3.599
7	7.186	6.809	5.420	5.212
8	9.084	8.560	6.455	6.177

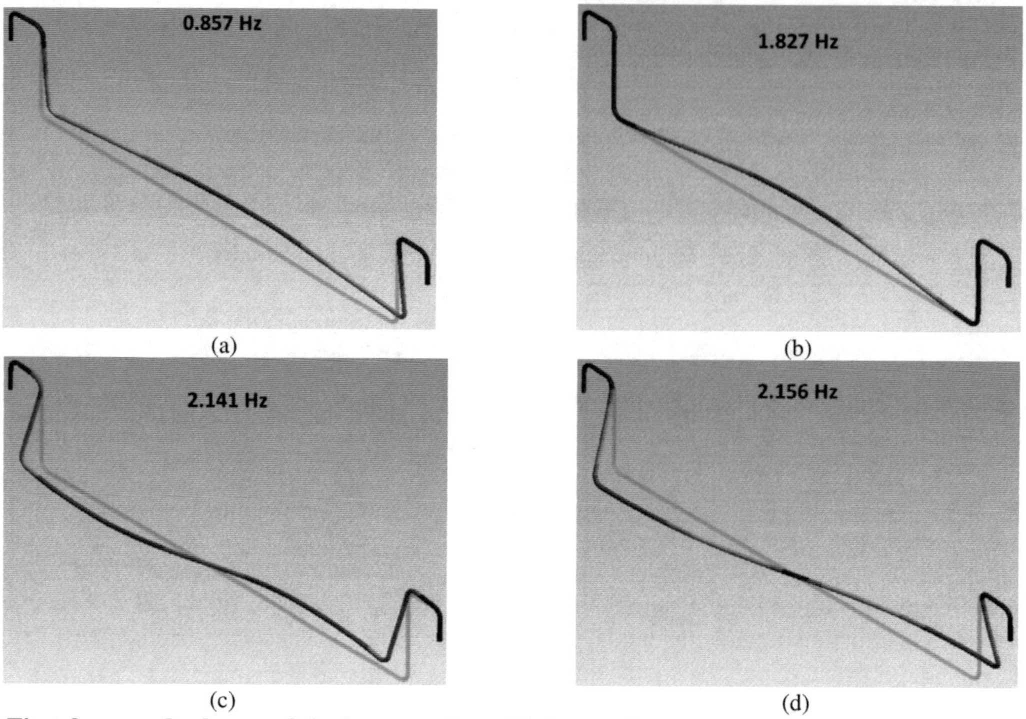

Figure 2: First four mode shapes of the jumper pipe with internal and external fluid: a) 0.856Hz; b) 1.827Hz; c) 2.142Hz; d) 2.156Hz.

CFD Results: Figure 3 shows the time history of oil flow into and out of the jumper pipe. The pulsating oil slugs were injected into the pipe at a frequency of 4.4 Hz, Figure 3-a. Figure 3-b shows the outlet oil mass flow rate and shows a clear transition point at 7.24 sec., where the oil flow rate becomes statistically stationary. Thus, we can estimate that the flow start up effects last for about 7.24 sec., i.e. until the oil flow becomes statistically stationary. The FFT of the outlet flow rate history, Figure 3-c, shows a dominant frequency at the slugging frequency, f_{slug}=4.4 Hz, which is a direct result of the slug injection frequency at the inlet. Additional frequencies can be seen at f_1=1.83 Hz and $f_2 = 0.5*f_{slug}$= 2.3 Hz. Although the primary interest in this study is to assess the pipe deformation and any resonant risk, the fluctuations of the oil flow could impact the hydrodynamic forces acting on the pipe and could in turn affect the deformation response frequency.

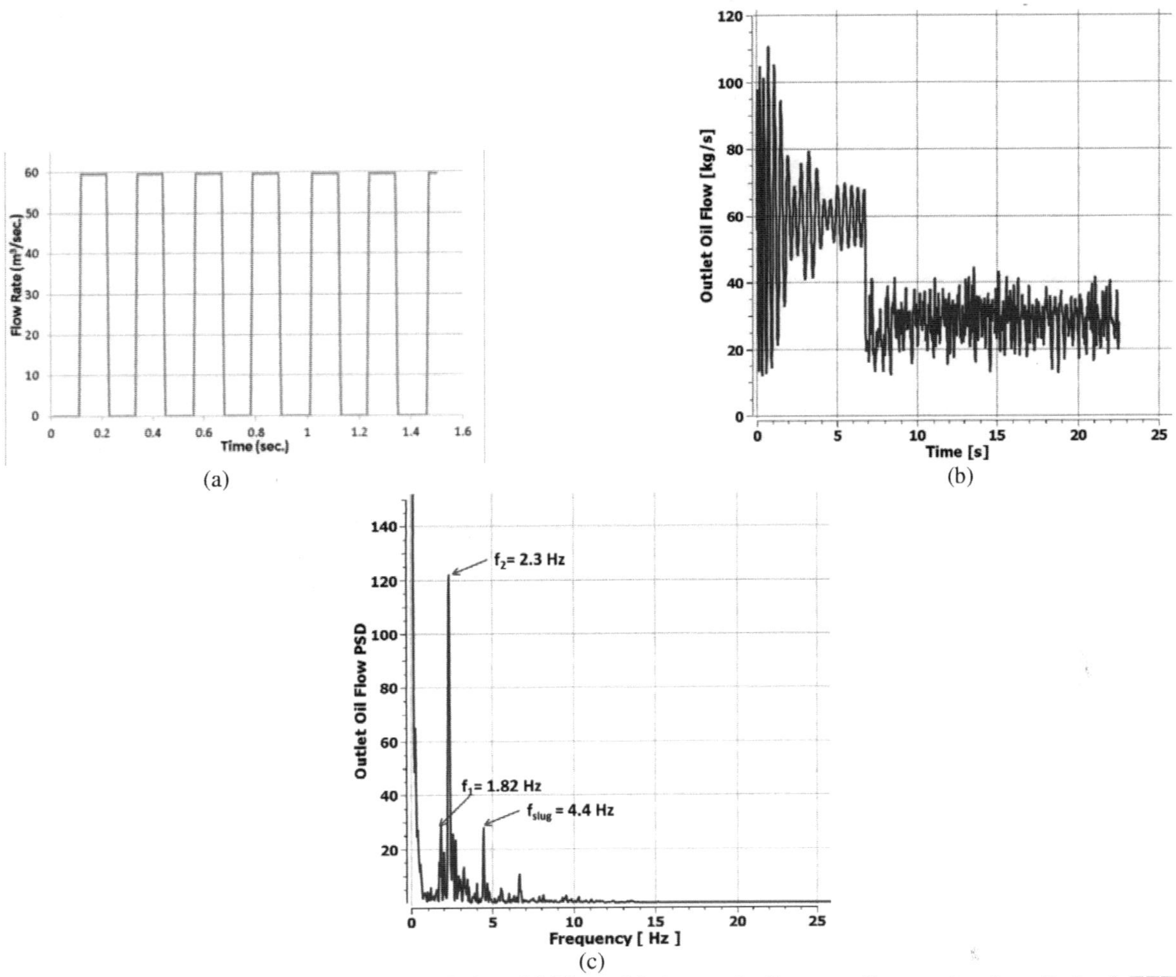

Figure 3: a) Pulsating oil flow rate at inlet; b) Time history of oil mass flow rate at outlet; c) FFT of oil mass flow at outlet.

Figure 4 shows the detailed multiphase distribution in the pipe at t=22.5 seconds at selected locations of the pipe. Although the injected oil slugs maintain their structure near the inlet, they start deforming as flow making turns, and the slugs are no longer distinguishable after the second flow turn, Figure 4-1. Moreover, stratified flow structure can be observed in the long horizontal span of the pipe due to the large density difference between oil and gas, Figure 4-3. Terrain induced slugging can be observed at bottom left bend and before exit bend of the pipe, Figure 4-2 and Figure 4-4. In general, the multiphase flow structure inside the pipe is fairly complex and exhibits different multiphase regimes, e.g. slug, annular, and stratified.

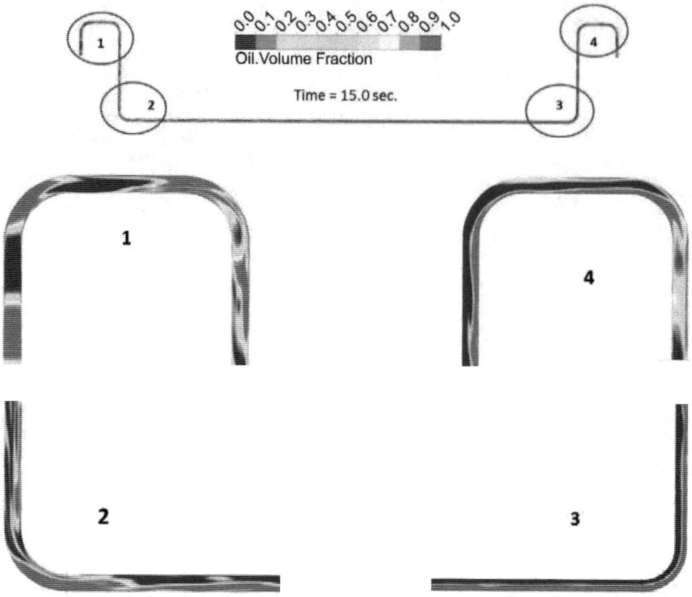

Figure 4: Case 1: Detailed Oil volume fraction distribution at pipe bends at t=22.5 sec.

Figure 5 shows the time history and FFT of the total deformations at selected point locations on the pipe; Elbow 1 (near inlet), Mid-span, and Elbow 2 (near outlet). The deformations history exhibits similar behavior at the different monitor locations in the domain, where it peaks at t=2.7sec with maximum deformation of 0.08m, then decays with time to maximum deformation of 1 mm, Figure 5-a. Figure 5-b shows the FFT analysis of the total deformation history, where maximum peaks are clearly visible at f_{slug} and its harmonics, $2f_{slug}$ and $3f_{slug}$. Moreover, a smaller peak occurs at 2 Hz. The detection of this frequency is consistent with the outlet flow rate discussed earlier. Under the conditions in this study, clearly there is a potential resonant concern around this frequency as it is very close to the second mode natural frequency of 1.725 Hz.

Figure 6 shows the total deformation of the pipe at selected time instances, t = 2.0, 4.0, and 16.0 seconds. As seen from the deformation shape, most of the deformation occurs in out of the pipe plane direction (normal to the page). Moreover, note that the pipe straight span experiences the maximum deformation in the pipe.

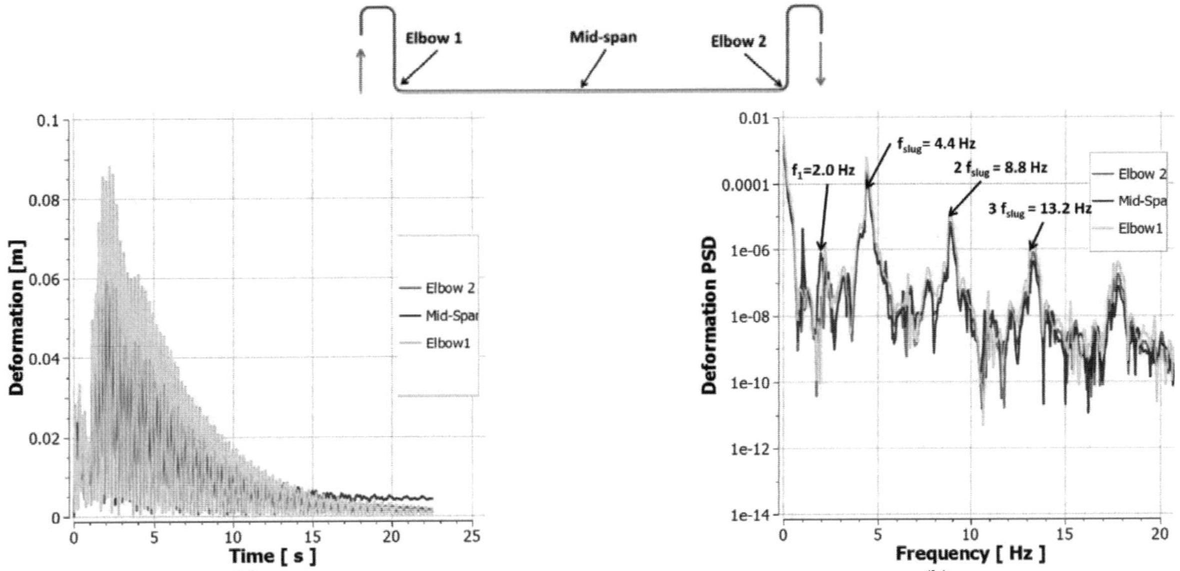

(a) (b)

Figure 5: structural total deformation at selected points: a) Time history; b) FFT

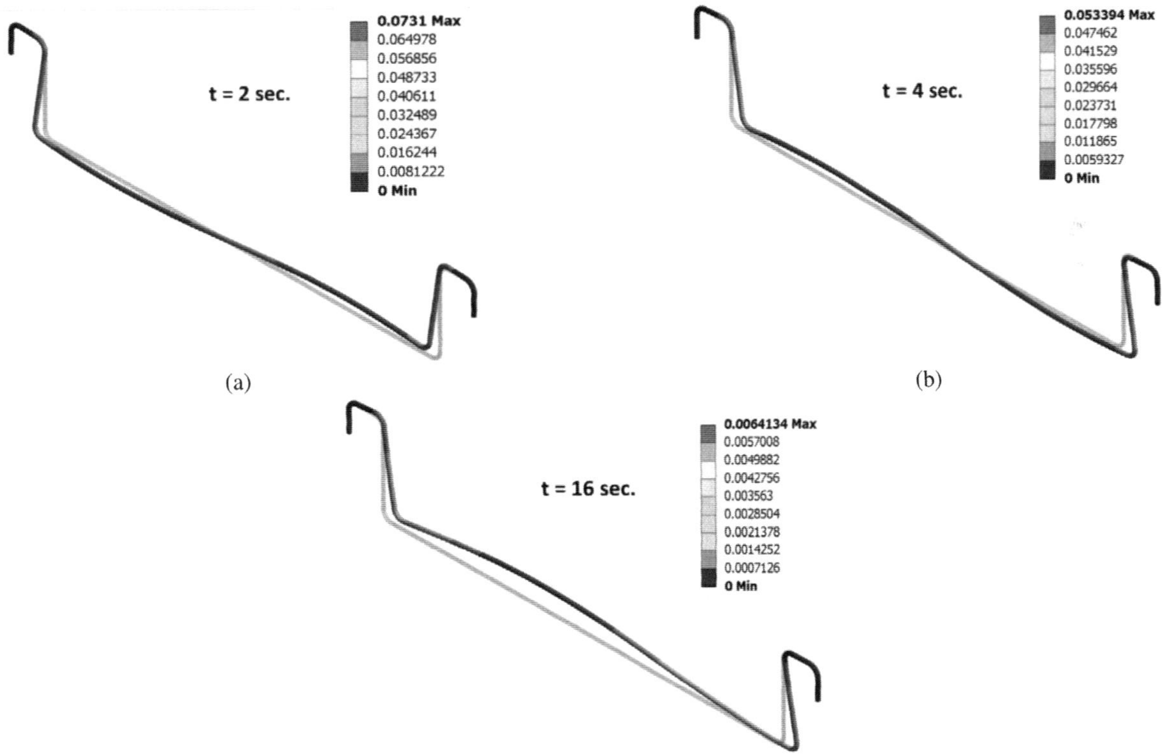

(a) (b) (c)

Figure 6: Jumper pipe total deformation at selected times: a) t =2.0 sec.; b) t=4.0 sec.; c) t=16.0 sec.

SUMMARY AND CONCLUSIONS

In this paper, FSI analysis of a submerged jumper pipe carrying multiphase flow during startup operation was simulated using fully coupled approach. Results show that

- Highest pipe failure risk occurs during the operation startup period, where the pipe is exposed to start-up conditions.
- The slug frequency is one of the main factors affecting the total fluid force acting on the pipe and the FIV response
- FSI analysis predict a key response frequency around 1.8 Hz. Given the conditions in this paper, this is a resonant risk as it is very close to the second mode natural frequency 1.725 Hz.
- During steady operation conditions, the maximum deformation is very small and weakly coupled FSI analysis should suffice.

REFERENCES

1. Swindell R., "Hidden Integrity Threat Looms in Subsea Pipework Vibrations." Off-shore Magazine, Sep. 2011.
2. Yih T.S., Griffith P., "Unsteady Momentum Fluxes in Two-Phase Flow and the Vibration on Nuclear Reactor Components." Report No. DSR 70318-58, Nov. 1968.
3. Sim W.G., Bae B.M., Mureithi N.W., "An Experimental Study on Characteristics of Two-Phase Flows in Vertical Pipe." J. Mechanical Science and Technology, vol. 24, no. 10, pp. 1981-1988. DOI: 10.1007/s12206-010-0706-8.
4. Riverin J.L., de Langre E., Pettigrew M.J., "Fluctuating Forces Caused by Internal Two-Phase Flow on Bends and Tees." Journal of Sound and Vibration, vol. 298, pp. 1088-1098, 2006.
5. API RP 1111, Recommended Practice for the Design, Construction, Operation, and Maintenance of Offshore Hydrocarbon Pipelines (Limit State Design), 3rd Edition, July 1999.
6. D. Jia, "Slug Flow Induced Vibration in a Pipeline Span, a Jumper, and a Riser Section." 2012 Offshore Technology Conference, paper no. OTC-22935.
7. D. Jia, "Effect of Buoyancy Conditions, Flow Rate, Slug Length, and Slug Frequency on Slug Flow Induced Vibration in a Pipeline Span." 2013 Offshore Technology Conference, paper no. OTC-24051.
8. Y. Lu, C. Liang, J.J. Manzano-Ruiz, K. Janardhanan, Y.Y. Perng, "FSI Analysis of Flow-Induce Vibration in Subsea Jumper Subject to Downstream Slug and Ocean Current." 2014 33rd Int. Conf. on Ocean, Offshore, and Arctic Engineering, OMAE2014-24120.
9. M. Elyyan, Y.Y. Perng, M. Doan, "Fluid-Structure Interaction of Subsea Jumper Pipe." 2014 33rd Int. Conf. on Ocean, Offshore, and Arctic Engineering, OMAE2014-24070.

400166 A Novel Simple Approach to Describe Sloshing in FPSO Vessel Using CFD Simulation

Abstract:

Liquid sloshing in vessels caused by forced acceleration has been the subject of intense investigations for last several decades both by experiments and numerical simulations. Many studies are done to minimize the sloshing induced forces on the vessel internals and some studies focused on different ways to describe the sloshing patterns. Most of the sloshing characterization methods are done using complex mathematical manipulation and more simplified method may be useful for better practical understanding. In this current study, simple/easily understandable methods are explored to describe sloshing phenomenon through Computation Fluid Dynamics (CFD) using ANSYS Fluent software in 3D cylindrical vessel. Several parameters including liquid level/tank length ratio, wave induced vessel motions due to acceleration, motion frequency, amplitudes in various sea state conditions (i.e. surge, sway, yaw, pitch, roll, heave). Parameters such as hydrodynamic force, pressure, moments, kinetic energy, potential energy, turbulent kinetic energy, height of the free surface, vorticity etc. are used to quantify the sloshing intensity. In addition, qualitative understanding of the sloshing motion is obtain through visual inspections of the gas-liquid/oil-water interface fluctuation, streamlines, velocity vector profiles etc. An equation connecting the independent variable to resultant quantities will be established. This method will make it easier to describe the sloshing state at various wave induced flow conditions.

400344 Attributes and Behaviours of Crude Oils That Naturally Inhibit Hydrate Plug Formation

Abstract:

The severe operating pressures and distances of deepwater tiebacks increase the risk of hydrate blockage during transient operations. In many cases, complete hydrate avoidance through chemical management may become cost prohibitive, particularly later in the field's life. However, for a unique subclass of crude oils that have not been observed to form a hydrate blockage during restart, active hydrate prevention may be unnecessary. Over the last 20 years, limited information has been reported about the chemical or physical mechanisms that enable this "magic" non-plugging behaviour. This paper demonstrates a systematic method of characterising whether an oil is "magic", including (i) physical property analysis that includes, and builds upon, ASTM standards; (ii) water-in-oil emulsion behaviour; and (iii) the effect of oil on hydrate blockage formation mechanics. This last set of experiments utilises a sapphire autoclave to allow direct observation of hydrate aggregation and deposition, combined with resistance-to-flow measurements. The effect of shut-in and restart on the oil's plugging tendency is also studied in these experiments. The method was tested with several Australian crude oils, some of which exhibited non-plugging behaviour. In general, these "magic" crude oils do not form stable water-in-oil emulsions, but do form stable, non-agglomerating hydrate-in-oil dispersions; the oils suppress hydrate formation rates and their resistance-to-flow does not increase significantly when the amount of hydrate present would normally form a plug. These results suggest that magic oils contain specialised, active chemistries that discriminate between hydrate-oil and water-oil interfaces.

401042 Reducing the Adhesion of Hydrates Using Durable Hydrate-Phobic CVD Polymers

Abstract:

The formation and accumulation of clathrate hydrate inside oil and gas pipelines causes severe problems in deep-sea oil and gas operations. Hydrate-phobic, durable, and mechanically-robust bilayer poly-divinyl benzene (pDVB)/poly(perfluorodecylacrylate) (pPFDA) coatings were developed using initiated chemical vapor deposition (iCVD) technique to inhibit the formation of hydrates and in the case of any formation reduce their adhesion to underlying substrates. Optical profilometer measurement was performed on the coating and root mean square (Rq) value of 18.9±5.4 nm was obtained. Mixtures of tetrahydrofuran (THF) and water were used to study the formation of hydrates and their adhesion strength. The weight percentage of the THF in water was changed from 0-70 and resulted in a decrease in surface tension of THF-water mixture from 70 to 28 mN/m. Goniometric measurements on coated substrates exhibited advancing water contact angle of 157.8° ± 2.3° and receding water contact angle of 131° ± 8°, whereas 70 weight % THF in water presented advancing angle of 85.1° ± 6.1° and receding angle of 48.5° ± 4.9°.

A 20 weight % THF in water mixture was found to be critical to form only hydrate when cooled down to low temperature at atmospheric pressure. The strength of hydrate adhesion experienced ten-fold reduction when substrates were coated with these iCVD polymers: from 1050±250 kPa on bare substrates to 128±100 kPa on coated ones. A weight % of THF in water lower than 20 resulted in formation of both ice and hydrate when cooled to low temperature. In this regime, a relation was obtained to express the adhesion strength of ice/hydrate mixture as a function of ice adhesion strength and hydrate adhesion strength. The adhesion strength of ice, hydrate, and ice/hydrate mixture were found to correlate well with the work of adhesion between liquid droplets used in their formation (water or THF-water mixture) and underlying substrate.

The weight % of THF in water increased above 20 and resulted in partial formation of hydrate with some unfrozen THF-water mixture. In this regime, the adhesion strength of hydrate was found to correlate well with the work of adhesion between 20 weight % THF in water droplets and underlying substrate. These data suggest that 20 weight % THF in water has to be used as a probe fluid to identify the strength of hydrate/unfrozen liquid adhesion to the substrates.

The impact of subcooling temperature and time on the adhesion strength of hydrate on coated and bare substrates was studied. These data suggest that the combination of time and subcooling temperature is important to form THF hydrates. However, once the hydrates formed, their adhesion strength were comparable regardless of their formation conditions. The results of this work suggest that THF-water mixture repelling nature of a given substrate can be utilized to assess its hydrate-phobic behavior; hence, it opens a new pathway for studying hydrate-phobicity.

399616 Avoiding Gas Hydrate Problems via Environmentally Friendly Gas Hydrate Inhibitors

Abstract:

One of the biggest problem in processing oil and natural gas resources is the often occurring blockage in the pipelines caused due to uncontrolled gas hydrate formation in the pipelines. Dehydration and the removal of the blockage safely by using chemical inhibitors causes great economic and operation time loss. There are several inhibitors that has been used to overcome this problem for several decades; however existing state of the art chemicals have their own operational problems that needs to be addressed. We aim to design and test novel hydrate inhibitors that are suitable for natural gas pipelines and its processing facilities and we are aiming to provide more effective and sustainable reservoir utilization and processing natrual gas reservoirs improving theflow assurance.

In this work, we present the initial findings of a Qatar Foundation (QF) funded project, which deals with the natural gas hydrate formation characteristics of Qatari type gas in both experimental (PVTx) and computational (molecular simulations) methods. We present the data from the two fully automated apparatus: a gas hydrate autoclave and a rocking cell. Hydrate equilibrium curves including growth/dissociation conditions for multi-component systems for several gas mixtures that represent Qatari type natural gas with and without the presence of well known kinetic and thermodynamic hydrate inhibitors. Choline Chloride, Choline Bistriflimide and Choline Acetate were studied in this work. These ionic liquids were used for testing their inhibition performance and their DFT and molecular modeling simulation results were also obtained and compared with the experimental results. They were used at 1% and 5% (wt) and at 5% amount, they showed good hydrate inhibition performance for both pure methane and gas mixtures as well.

400352 A New Application of Differential Scanning Calorimetry to Identify the Adsorption of Surfactants on Gas Hydrate Interfaces

Abstract:

Gas hydrates represent a critical risk in subsea flow assurance, where the aggregation or deposition of hydrate particles in the flowline may result in complete flowline blockage. In conventional designs, gas hydrates are avoided altogether by injecting thermodynamic chemical inhibitors, such as methanol. In the past decade, hydrate anti-agglomerants (AAs) have been developed to enable a new generation of subsea system design, where the risk of hydrate blockage is managed inside the hydrate stability region. The performance of each AA depends on both the chemistry and physical properties of the aqueous and hydrocarbon phases, which may traditionally be validated by qualitative autoclave or rocking cell tests. We present a new strategy for quantifying the stability of a hydrate-in-oil dispersion with and without AAs present, based on the use of a differential scanning calorimeter (DSC). The DSC cells are filled with a water-in-crude oil emulsion and pressurised to 1000 psi; cell temperature is cycled 10 times between 20 and -40 °C to repeatedly form and dissociate gas hydrate. As the dispersion destabilises, the amount of hydrate formed in each subsequent cycle decreases. This behaviour is quantified by a change in total heat flow, which is measured by the DSC. We have applied this measurement strategy to a water-in-crude oil emulsion containing representative industrial surfactants over a wide range of concentration. The DSC data provides evidence to quantify the maximum dispersion stability gained through the surfactant, as well as a relationship between the dispersion stability and chemical concentration. The minimum adsorption concentrations for the surfactant was successfully identified and compared to previously-reported estimates for hydrate systems. The results indicate that simple quaternary ammonium surfactants are active at the hydrate-oil interface at mass fractions above 0.001% in oil. A secondary suite of tests was performed to quantify the effect of brine salinity on dispersion stability. As the mass fraction of NaCl in water approached 10 wt%, the hydrate dispersion became stable; this result suggests the importance of using representative aqueous phase samples when evaluating chemical performance.

399620 Formation of Hydrate Plug in Water-Decane-Natural Gas Systems with Varying Watercut and Inhibitor Concentrations

Abstract:

The risk of hydrate blockage in subsea oil and gas pipelines represents a critical operational and safety hazard. To date, many studies have focused on predicting hydrate growth and plug formation in oil-continuous systems at low watercut. This study used a high-pressure autoclave to investigate hydrate growth rate and resistance-to-flow at watercuts of 20 to 100% in decane. The results suggest that the high watercuts generated late in the field life do not represent the greatest plugging risk. Instead, moderate watercuts of 40-60% show the largest increase in resistance-to-flow (motor torque), with deposition and sloughing-type behavior apparent in the signal. The motor torque increases associated with this high-risk region were effectively suppressed with the addition of both 10 and 30 wt% MEG in the aqueous phase, where 42 wt% MEG was required for complete inhibition in the systems studied. In the 10 wt% MEG cases, the relative torque increased smoothly from 1 to 2 or 2.5 N·cm at both watercuts. These mild torque increases were fully suppressed through the addition of 0.5 wt% PVCap in the oil phase. No torque increases were detected for the 5 vol% hydrate formed in the 30 wt% MEG trials. Together, these data suggest that mild under-inhibition (within 25% of the required quantity) could lower the risk of hydrate blockage sufficiently to warrant flow loop or even field-scale trials.

399673 Probing Asphaltene Deposition Using Microfluidic Devices

Abstract:

Asphaltene deposition is a common cause of significant flow assurance problems in wellbores and production equipment as well as near-wellbore regions in oil reservoirs. Therefore, a better understanding of the mechanism by which asphaltenes deposit onto solid surfaces is highly desirable. Here we proposed a novel methodology for probing asphaltene deposition using microfluidic devices that model porous media. A variety of factors influencing asphaltene deposition will be investigated, including fluid flow pattern, asphaltene polydispersity, and solvent effects. Real time visualization of asphaltene deposition in microfluidic channels is realized. First, this micromodel is well characterized with known properties of flow in the oilfield. Second, the model system of pure C5+ asphaltenes in toluene with a known asphaltene instantaneous onset concentration is used to eliminate the disturbance from unknown components in crude oil. Coupled with the indirect method and images of aggregates, precipitation and aggregation effects are discussed with deposition rate by varying the injection ratio of n-heptane in the porous media both in the pore scale (visualization) and macroscopically qualitatively (collector efficiency). By understanding the nature and the mechanisms of asphaltene deposition, we increase our ability to design cost effective mitigation strategies that includes the development of a new generation of asphaltene deposition inhibitors and improved methods for prevention and treatment of this problem.

398122 Evaluating Effect of Compositional Gradients on Asphaltene Stability in Reservoirs with CPA and PC-SAFT

Abstract:

Optimal field development requires proper knowledge of the composition and phase behaviour of the fluids that are present in the reservoir. Within the complex matrix of reservoirs, various forces such as gravitational and thermal gradient induced forces as well as mass flux of light hydrocarbons at reservoir boundaries can cause the migration and compositional distribution among the different reservoir units or compartments, affecting optimization and management of hydrocarbon production. For an asphaltenic reservoir field, the large compositional variations may lead to asphaltene precipitation and deposition within reservoirs, causing formation damage and permeability reduction. Taking into account all the phenomena, to establish a consistent picture of fluid distribution and phase behaviour in the field relies on the fundamental understanding of the effects of compositional variations on fluid phase behaviour.

In this study our well-developed Cubic-Plus Association (MF-CPA EoS) model[1] and the perturbed chain-statistical associating fluid theory equation of state (PC-SAFT EoS)[2,3] are applied to model the effect of compositional gradients as well as changes of pressure and temperature on asphaltene stability in reservoirs. Possible size distribution of the asphaltene monomers based on Gamma distribution is also considered to assess the impact of the size distribution on the phase behaviour of asphaltenes in various reservoir fluids.

Real field cases with compositional gradients, and severe asphaltene precipitation and deposition problems from a wide range of asphaltenic crudes ranging from heavy oils to light live crudes were used to validate the models.

The validation and comparison demonstrate the capabilities and advantages of both the conventional cubic equation of state plus association and PC-SAFT model. With its simplicity, capabilities and consistency with the industry standard fluid characterisation methods, MF-CPA EoS has the potential to become a promising and general engineering tool for providing essential knowledge of asphaltene phase behaviour in reservoirs.

[1] Xiaohong Zhang, Nuno Pedrosa, and Tony Moorwood, Energy Fuels, 2012, 26, 2611−2620.

[2] Francisco M. Vargas, Jeff L. Creek and Walter G. Chapman, Energy Fuels, 2010, 24 (4), pp 2294–2299.

[3] Gonzalez, D. L.; Hirasaki, G. J. Creek, J.; Chapman, W. G. Energy & Fuels 2007, 21, 1231-1242.

400354 Cohesive Force Measurement Between Precipitated Asphaltene Solids

Abstract:

Asphaltene solids may precipitate in subsea crude oil wellbores and flowlines, due to the rapid decrease in fluid pressure during flow. While the consequence of asphaltene deposition and build-up at the wall is well characterised, little is known about how precipitated asphaltene particles behave in flow. If secondary solids are also present, asphaltene particles with a strong aggregation potential may interact with other solids, such as gas hydrate. In this presentation, we discuss adaptations to a micromechanical force (MMF) apparatus that enable measurement of cohesive force between asphaltene particles, which were precipitated from crude oil diluted by various light aliphatic hydrocarbons and filtered on a 0.45-micron nylon paper. The results demonstrate that asphaltene cohesive forces may be up to twice those measured previously for gas hydrate, with solid-solid cohesion as the primary mechanism. MMF data provides further insight into the asphaltene particle surface free energy. The cohesive force data demonstrate that this free energy is very sensitive to the type of crude oil from which the asphaltenes were extracted, suggesting the potential for MMF-based studies to characterise the inherent activity of the asphaltene species within crude oil. For three different Australian crude oils, further tests were performed after the asphaltene-active binding resins were stripped from the asphaltene surface. The data demonstrates a uniform shift toward higher cohesive force, suggesting that these active resins may function to decrease asphaltene surface free energy.

401693 Interfacial Properties of Asphaltenes at the Heptol-Brine Interface

Abstract:

Water-oil emulsion is inevitable in the production of crude oil. Understanding the role of asphaltenes at the heptol (mixture of *n*-heptane and toluene)-brine interface is of significant importance to deal with oil lost and the waste disposal. A Tracker tensiometer was used to test the influence of asphaltenes concentration, time, temperature (25.0-85.0) and large inorganic particles on the interfacial tension of heptol-brine interface. A transparent quartz packed-bed microreactor with in-line analytics was designed to test the emulsion block in porous media. Experimental results show that the presence of asphaltenes and large hydrophobic inorganic particles in heptol could stabilize the water in oil emulsions. The results could aid in the design of remediation to improve water displacement of crude oil in sandstone reservoirs and enhance the crude oil recovery.

Effect of shear history on restart of gelled waxy crude oil pipeline

Youquan Bao, Jinjun Zhang

National Engineering Laboratory for Pipeline Safety/Beijing Key Laboratory of Urban Oil & Gas Distribution Technology, China University of Petroleum, Beijing 102249, China

Abstract: The existence of wax in waxy crude oil makes its rheological property very complex. When the temperature of waxy crude oil is high enough, wax dissolves in the oil in molecular form and the whole system behaves as Newtonian fluid. When the temperature is reduced to a certain value which is called "wax appearance temperature", the dissolved wax begins to precipitate from the oil as a result of supersaturated and the whole system is still Newtonian fluid. If the temperature is further reduced, the amount of precipitated waxy will increases gradually, and the system turns from Newtonian to non-Newtonian. Once the mass percent of the precipitated waxy increases to 1wt%~6wt%, the wax crystals in crude oil will attract and interlock with each other to form a three-dimensional sponge-like network structure where liquid oil is wrapped in, and the whole system turns from sol to gel. The gelled waxy crude oil shows complex rheological behaviors, such as viscoelasticity, yield stress, and thixotropy.

Waxy crude oil is always heated and transported by pipeline for its bad fluidity at normal temperature. In the production process, shutdown of pipeline is unavoidable. If the shutdown costs a long time, the waxy crude oil in the pipeline will gelled because of temperature reduction. How to restart the pipeline successfully is one of the great concerns of flow assurance, and it is also a research hotspot in the last few decades.

Shear history has greatly influence on rheological properties of waxy crude oil, while the rheological properties have a determinative effect on pipeline restart. In the present work, shear history was simplistically simulated by means of rheometer (HAAKE MARS III) firstly, and the main processes include: dynamically cooling (shear rate is 10 s^{-1}/50 s^{-1}/100 s^{-1}) the waxy crude oil sample in the rheometer from 50 °C to 34 °C at a cooling rate of 0.1 °C/min; Statically cooling the sample further to 32 °C. Then, shear rate stepwise increasing experiment was conducted to investigate the effect of shear history on the structure breakdown characteristic of waxy crude oil, and Eq. (1) was employed to describe the breakdown characteristic. Finally, numerically simulation was performed to study the effect of shear history on the restart process.

$$\begin{cases} \tau = \tau_y + (K + \lambda \Delta K)\dot{\gamma}^{n_1} \\ \dfrac{d\lambda}{dt} = \dfrac{1}{1+\gamma^{n_2}}[a(1-\lambda) - b\lambda\dot{\gamma}^{m}] \end{cases} \quad (1)$$

Where is shear stress; is shear rate; is structural parameter, and; is total shear strain;, *K*, , n_1, n_2, *a*, *b* and *m* are undetermined parameters.

τ

$\dot{\gamma}$

λ

$\lambda \in [0,1]$

γ

τ_y

ΔK

The shear rate stepwise increasing results under different shear history were shown in Fig. 1.It can be seen from Fig.1 that with increasing shear rate during the cooling process, the stress overshoot increases and the breakdown curve is upward. The Eq. (1) was used to quantify the curves in Fig.1 and the associated parameter values were given in Table 1.

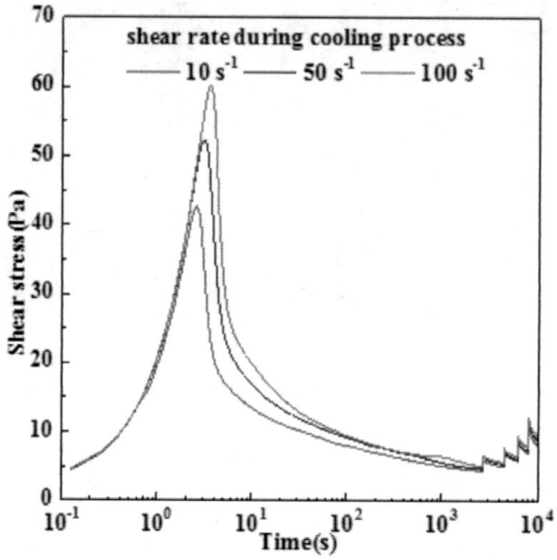

Fig. 1 Shear rate stepwise increasing results under different shear history

Table 1 Associated parameter values of Eq. (1) under different shear history

Shear rate during cooling process/s^{-1}	τ_y Pa	K $Pa \cdot s^{n_1}$	ΔK $Pa \cdot s^{n_2}$	n_1 -	n_2 -	a -	b -	m -
10	24.74	0.67	15.63	0.6912	0.8691	0	0.0123	1.1014
50	27.87	0.97	23.93	0.6606	0.8580	0	0.0151	0.9847
100	28.90	1.15	31.26	0.6298	0.8352	0	0.0210	0.8949

Considering a horizontal pipe of length L=1 km and inner diameter d=0.5 m. The weakly compressible () fully-gelled waxy crude oil is assumed to have

$\alpha = 5\times10^{-9}\text{Pa}^{-1}$

uniform density and rheological properties (as shown in Table 1) initially. A constant pressure gradient MPa is suddenly applied at the pipe inlet to break

$\rho = 850\ \text{kg/m}^3$

$\Delta P = 1$

the gel and restart the flow. The results of numerical calculation were normalized according to the following parameters:

$$P^* = \frac{P}{\Delta P},$$

$$V^* = \frac{V}{\Delta P d^2/(32\mu_{ref}L)}, (\mu_{ref} = 1\ \text{Pa}\cdot\text{s})$$

$$,t^* = \frac{t}{L/\sqrt{\rho\alpha}},$$

$$z^* = \frac{z}{L}.$$

The inlet and outlet velocities variations with time for different cases were shown in Fig. 2. It can be seen from Fig. 2 that, for all three case, inlet velocities increase instantaneously at t^*=0 while outlet velocities increase at t^*=1 and both oscillate before reaching the steady states. In addition, the higher the shear rate during cooling process, the higher the steady inlet/outlet velocity is.

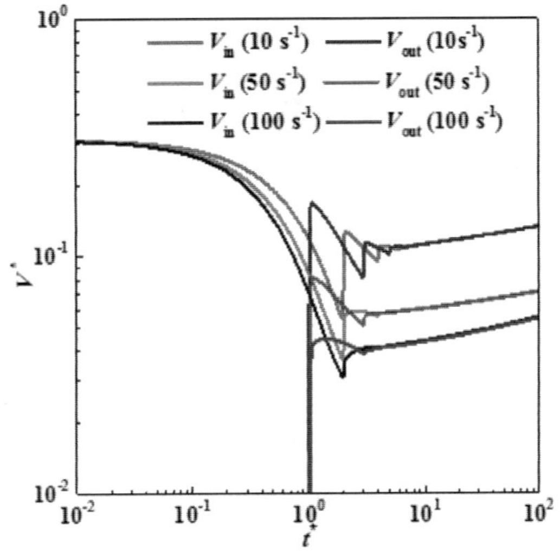

Fig.2 Time variations of inlet and outlet velocities for different cases

The time variations of pressure at Z^*=0.5was shown in Fig. 3. It can be concluded that lower shear rate during cooling process results in higher pressure-peak magnitude. However, the steady-state pressure at Z^*=0.5 is independent of the shear history.

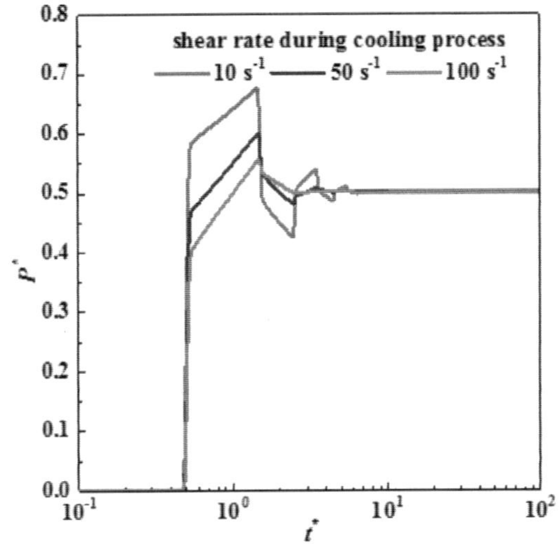

Fig. 3 Time variations of pressure at Z^*=0.5 for different cases

Key words: waxy crude oil; shear history; restart

402009 Evaluating Wax Gelation Risk for Restart Operations - Closing the Gap between Academic Research and Industrial Applications

Abstract:

Evaluating Wax Gelation Risk for Restart Operations – Closing the Gap between Academic Research and Industrial Applications

Zhenyu Huang, Assured Flow Solutions LLC

The problem wax gelation is well known to the flow assurance industry. A waxy crude oil can form gels of high viscosity and yield stress during shut-down, which poses significant concern for the pressure requirement during field restart. Despite recent advances in academic research to study this problem, the industry still relies heavily on conventional fluid testing and calculation methods to tackle wax gelation, which often leads to over-conservative mitigation strategies.

With the aim to resolve this disconnection, this presentation discusses the importance of understanding the complex rheological characteristics of the waxy gel in order to select appropriate experimental measurements and correctly interpret the results. Conventional rheological testing methods can be over-conservative, while more advanced characterization methods still need to be applied/analyzed with prudence.

In addition, restart modeling is an integral part to develop rigorous understanding of the pressure requirement in the field. In this presentation, the behavior of a waxy gel during restart will be discussed and several existing models will be carefully reviewed. Finally, the application of an in-house restart model to identify the restart challenges in the field is presented.

401871 Mining Big Data from Product & Process Simulations to Improve Designs

Abstract:

Advancements in computer modeling & simulation, also known as virtual prototyping, have already dramatically reduced the need for exhaustive physical prototyping and testing of complex subsea subsystem, while concomitantly increasing the understanding of the proper and safe operating ranges of such equipment. While these advancements have provided a significant improvement in speed-to-market and an associated reduction in testing costs, there remains a huge untapped opportunity to also reap major product design improvements by better leveraging the potential of virtual design exploration.

In this paper, we will demonstrate how it is now possible to leverage three available technologies to mine massive simulation data and discover better subsurface and subsea designs, *faster* than ever before. The combination of robust computer-aided engineering simulation models, coupled with ubiquitous high performance computing resources, and innovative direct design search strategies enable radical improvement in the quest for designs that perform better, cost less, and operate more safely for a longer period of time. We will demonstrate the effectiveness of this big data mining strategy in the design of subsea heat exchangers, off-shore risers, and oil drilling equipment.

401670 Building Sustainable Data Quality Programs for E&p : Best Practices

Abstract:

Data Quality is a significant barrier at O & G organizations in adopting Business Intelligence platforms.

Enterprise Data Management in O & G companies can improve and sustain Data Quality leading to more agile decision making across the enterprise.

The paper discusses the best practices in data management and charts a roadmap for upstream O & G organizations to improve their data quality in a sustainable way.

This is substantiated by two Case Studies.

399978 A Distributed Modeling Framework for Monitoring Big Data in Plant-Wide Processes

Abstract:

With the growing complexity of the modern process industry, the era of big data has arrived, particularly for the large scale plant-wide processes. The size of data can range from terabyte to petabyte in the plant-wide process even for a single dataset. This big size of data poses great challenges not only for information capture, data management and storage, but also more important, to efficiently interpret the information hidden within those data. In this paper, a big data oriented distributed modeling framework is developed for plant-wide process monitoring. Based on this framework, the whole plant-wide process is decomposed into different blocks, and statistical data models are constructed inside those blocks. For online monitoring, results obtained from different blocks are integrated through the decision fusion algorithm. A detailed case study is carried out for performance evaluation of the plant-wide monitoring method. Research challenges and perspectives are also discussed and highlighted for future work.

AUTHOR INDEX

Aichele, Clint P. 68
Akella, Venkata 94
Albrecht, Connor R. 19, 37
Aman, Zachary 84
Aman, Zachary M. 25, 26, 80, 83, 87
Aparicio, Santiago 82
Atilhan, Mert 82
Bao, Youquan 89
Barsotti, Domenic 93
Biswal, Sibani Lisa 24, 85
Boggara, Mohan 22
Bonneville, Alain 29
Carroll, Kenneth 29
Chai, John 85
Chan, Lester Lik Teck 95
Chen, Chau-Chyun 18, 20, 27
Chen, Junghui 95
Clark, Peter 31
Correa, Sarah 21
Cox, Quinton 21
Daniel, Alden 68
Dennis, Brian 7
Deodhar, Subodh 4
Di Lorenzo, Mauricio 25
Dong, Zhao 30
Duran-Sesin, Rodrigo 2
Elyyan, Mohammad A. 69
Esch, Thomas 4
Fatt, Yap Yit 85
Fernandez, Carlos A. 29
Ferramosca, Adrian 30
Flemings, Angelo 93
Garcia, Caitlin 21
Garcia, Miguel 21, 22
Ge, Zhiqiang 95
Gilmer, Matthew W. 23
Gleason, Karen 81
Goharzadeh, Afshin 85
Graham, Brendan F. 80, 87
Haber, Agnes 80, 83
Han, Shanpeng 14
Hartman, Ryan L. 88
He, Peng 24, 85
Heldebrant, David J. 29
Hoyt, David 29
Hu, Chuntian 88
Huang, Zhenyu 92
Ike, Shirley 23
Islam, M. R. 27
Ji, Sun Young 85
Johns, Michael L. 25, 26, 80, 83, 87
Jung, Gonghyun 28
Kabilan, Senthil 29
Khan, Rafiqul 3, 79
Kim, Jakyung 84
Koteeswaran, Samyukta 31
Kouba, Gene 68
Kozielski, Karen 25
Kraume, Matthias 63
Kumar, Neeraj 42
Lal, Manish K. 67
Lavenson, David 12, 68
Lawal, Akanni S. 19, 37, 38, 42, 46, 48, 64, 67
Li, Hongying 14
Lin, Yu-Jiun 24, 85
Maass, Sebastian 63
Mahmud, Md. A 3, 79
Masuku, Cornelius Mduduzi 49
Mathew, Nevin Thunduvila 85
May, Eric F. 84, 25, 26, 80, 83, 87
McKinley, Gareth H. 81
Melendez, Ariana 21
Mohammad, Sayeed 68
Mohan, L. Srinivasa 4
Morrissy, Shane A. 87
Mukherjee, Amrita 50
Mustafa, Merih 86
Najafi, Ali 69
Norris, Bruce W. E. 25, 26
Ortiz, Aldo 67
Panacharoensawad, Ekarit 13
Pandya, Deval 7
Pedrosa, Nuno 86
Perng, Yeong Yan 69
Pickering, Paul 80
Pisupati, Sarma 50
Read, Alex 93
Rittirong, Ake 13
Roehner, Richard 6
Rose, Kelly 2
Russell, Ronnie 7
Sambath, Krishnaraj 12
Sarica, Cem 13
Schlanger, Marvin O. 1
Seo, Yutaek 84
Shankar, Santhosh 28
Shao, Hongbo 29
Sharma, Kal Renganathan 17
Sim, Lawrence 2
Sisco, Caleb 22
Sohn, Young Hoon 84
Sojoudi, Hossein 81
Srinivasan, Vedanth 4
Subramani, Hariprasad J. 68
Syddall, Bill 80
Szczepanski, Richard 30, 86
Tanveer, Sheik 20
Tavakkoli, Mohammad 21, 85
Tododo, Akindele 42
Tohidi, Bahman 82

AUTHOR INDEX

Tsai, Kuochen....5
Ulrich, Stefan....63
Um, Wooyong....29
Umhoefer, Joe....2
Varga, Tamas....29
Vargas, Francisco....21, 22, 24, 85
Venkatesan, Rama....12
Vielma, Jason....2
Vogt, Sarah J.....83
Walsh, Matthew....81
Wang, Meng....27
Wang, Xiaolong....14
Whiteley, James R.....68
Xu, Fabin....30
Xu, Qiang....3, 79
Yen, Andrew....88
Yu, Bo....14
Yu, Yue....18
Zeng, Chunmin....30
Zhang, Chong....30
Zhang, Jinjun....14, 89
Zhang, Xiaohong....30, 86
Zhong, Lirong....29
Zou, Kaijuan....64